Discovering Genomics Proteomics And Bioinformatics

Discovering Genomics Proteomics And Bioinformatics

Dr. Jitendra Singh

MANGLAM

MANGLAM PUBLICATIONS

DELHI-110 053 (INDIA)

Published by :
MANGLAM PUBLICATIONS
L-21/1, Street No. 5, Shivaji Marg, Near Kali Mandir,
J.P. Nagar, Kartar Nagar, West Ghonda, Delhi-110053
Phone : 011-22945677, (M.) 9968367559, 9868572512
E-mail: manglam.books2007@rediffmail.com
manglam.publications2007@rediffmail.com

Discovering Genomics, Proteomics And Bioinformatics

First Edition 2009

ISBN 978-93-80013-04-6

PRINTED IN INDIA

Published by Ravindra Kumar Yadav for Manglam Publications, Delhi-110053 Laser Typeset by M. K. Computers, Delhi- 110 053, Printed at Sachin Printers, Moujpur Delhi- 53

PREFACE

The pace of progress in the branch of Genomics, Proteomics and Bioinformatics is fast and furious, particularly since the advent of genetic engineering and the recent advances in computer science and process control. The book addresses the multidisciplinary nature and the many fascinating aspects of genomics and proteomics thus providing a stepping stone in its progress as we enter a new era in which the use of computer recognized as an urgent need. The present title has been written for the undergraduate and post graduate students of Biological science. The title is not intended to be comprehensive nor could it be at length, but it concentrates as putting across the basic principles of the subject as briefly and lucidly as possible. It does this with the aid of easy and understandable illustrations. The aim is to enthuse the reader with this active and exciting area of research and to lay a solid foundation on which further study of its various facets may be based.

The author expresses his thanks to all those friends, colleagues, and research scholars whose continuous inspirations have initiated him to bring this title.

The author wishes to thank the M/s. Manglam Publications, Delhi for bringing out this book.

Constructive criticisms and suggestions for improvement of the book will be thankfully acknowledged.

Author

Contents

1

Introduction

A genome sequence is not an end in itself. A major challenge still has to be met in understanding what the genome contains and how the genome functions. The former is addressed by a combination of computer analysis and experimentation with the primary aim of locating the genes and their control regions and assigning a role to each gene. The bulk of this chapter will be devoted to these methods. The second question, understanding how the genome functions is, to a certain extent, merely a different way of stating the objectives of molecular biology over the last 30 years. The difference is that in the past attention has been directed at the expression pathways for individual genes, with groups of genes being considered only when the expression of one gene is linked to that of another. Now the issues have become more general and relate to the expression of the genome as a whole.

1.1 LOCATING GENES IN DNA SEQUENCES

Once a DNA sequence has been obtained, whether it is the sequence of a single cloned fragment or of an entire chromosome, then various methods can be employed to locate the genes that are present. These methods can be divided into those that involve simply inspecting the sequence, by eye or more frequently by computer, to look for the special sequence features associated with genes, and those methods that locate genes by experimental analysis of the DNA sequence.

1.1.1 Gene Location by Sequence Inspection

Sequence inspection can be used to locate genes because genes are not random series of nucleotides but instead have distinctive features. These features determine whether a sequence is a gene or not, and so by definition are not possessed by noncoding DNA. If in the future we can fully understand the exact nature of the specific sequence features that define a gene then sequence inspection will become a foolproof way of locating genes. We are not yet at this stage but sequence inspection is still a powerful tool in genome analysis.

1.1.1.1 Genes are open reading frames

Genes that code for proteins comprise *open reading frames* (ORFs) consisting of a series of *codons* that specify the amino acid sequence of the protein that the gene codes for. The ORF begins with an *initiation codon,* usually but not always ATG, and ends with a *termination codon,* either TAA, TAG or TGA: Searching a DNA sequence for ORFs that begin with an ATG and end with a termination triplet is therefore one way of looking for genes. The analysis is complicated by the fact that each DNA sequence has six *reading frames,* three in one direction and three in the reverse direction on the complementary strand, but computers are quite capable of scanning all six reading frames for ORFs. How effective is this as a means of gene location?

The key to the success of ORF scanning is the frequency with which termination triplets appear in the DNA sequence. If the DNA has random sequence and a GC content of 50% then each of the three termination triplets – TAA, TAG and TGA – will appear, on average, once every 43 = 64 bp. If the GC content is >50% then the termination triplets, being AT-rich, will occur less frequently but one would still be expected every 100-200bp. As there are three termination triplets and three reading frames in either direction, random DNA should not show many ORFs longer than 50 triplets in length, especially if the presence of a starting ATG is used as part of the definition of an 'ORF'. Most genes, on the other hand, are longer than 50 codons (the average lengths are 317 codons for *Escherichia coli* and 483 codons for *Saccha romyces cerevisiae*). ORF scanning, in its simplest form, therefore takes a figure of, say, 100 codons as the shortest length of a putative gene and records positive hits for all ORFs longer than this.

How well does this strategy work in practice? With bacterial genomes, simple ORF scanning is an effective way of locating most of the genes in a DNA sequence. This is illustrated by *Figure,* which shows a segment of the E. *coli* genome with all ORFs longer than 50

```
        GAS→
     TGA →
  ATG→
  5'-ATGACGAGAGAGCAGCCA TTTT AG-3'
  3'- T ACTGCTCTCTCGTCGGT AAAA TC-5'
                                          ← ATC
                                       ← AAT
                                    ← AAA
```

Figure 1.1. A double-stranded DNA molecule has six reading frames.

triplets highlighted. The real genes in the sequence cannot be mistaken because they are much longer than 50 codons in length. With bacteria the analysis is further simplified by the fact that there is relatively little noncoding DNA in the genome (only 11% for *E. coli*). If we assume that the real genes do not overlap, and there are no genes-within-genes, which are valid assumptions for bacterial DNA, then it is only in the noncoding regions that there is a possibility of mistaking a short, spurious ORF for a real gene. So if the noncoding component of a genome is small then there is a reduced chance of making mistakes in interpreting the results of a simple ORF scan.

1.1.1.2 Simple ORF scans are less effective with higher eukaryotic DNA

Although ORF scans work well for simple bacterial genomes, they are less effective for locating genes in DNA sequences from higher eukaryotes. This is partly because there is substantially more space between real genes (70% of the human genome is intergenic), increasing the chances of finding spurious ORFs. But the main problem with the human genome and those of higher eukaryotes in general is that their genes are often split by introns and so do not appear as continuous ORFs in the DNA sequence. Many exons are shorter than 100 codons, some less than 50 codons, and continuing the reading frame into an intron usually leads to a termination sequence that appears to close the ORF (*Figure*). In other words, the genes of higher eukaryotes do not appear in their DNA sequences as long ORFs and simple ORF scanning cannot locate them.

Solving the problem posed by introns is the main challenge for bioinformaticists writing new software programs for ORF location. Three modifications to the basic procedure for ORF scanning have been adopted:

1. *Codon bias* is taken into account. 'Codon bias' refers to the fact that not all codons are used equally frequently in the genes of a particular organism. For example, in human genes the

amino acid alanine is only infrequently specified by the codon GCG, usually one of the other three codons for alanine (GCA, GCC or GCT) is used. Similarly, threonine is most frequently coded by ACA, ACC or ACT and less often by ACG. The biological reason for codon bias is not understood, but all organisms have a bias, different in different species. Real exons are expected to display the codon bias but chance series of triplets do not. The codon bias of the organism being studied is therefore written into the ORF scanning software.

2. *Exon-intron boundaries* can be searched for as these have distinctive sequence features, though unfortunately the distinctiveness of these sequences is not so great as to make their location a trivial task. The sequence of the upstream, exon-intron boundary is usually described as

   5' -AG↓GTAAGT -3'

   the arrow indicating the precise boundary point. However, only the 'GT' immediately after the arrow is invariable, at the preceding positions nucleotides other than the ones shown are quite often found. In other words, the sequence shown is a consensus, the average of a range of variabilities. The downstream, intron-exon boundary is even less well defined:

   5' -PyPyPyPyPyPyNCAG↓-3'

   where 'Py' means one of the pyrimidine nucleotides, T or C, and 'N' is any nucleotide. Simply searching for the consensus sequences will not locate more than a few exon-intron boundaries because most have sequences other than the ones shown. Writing software to take account of the known variabilities has proven difficult, and at present locating exon-intron boundaries by sequence analysis is a hit and miss affair.

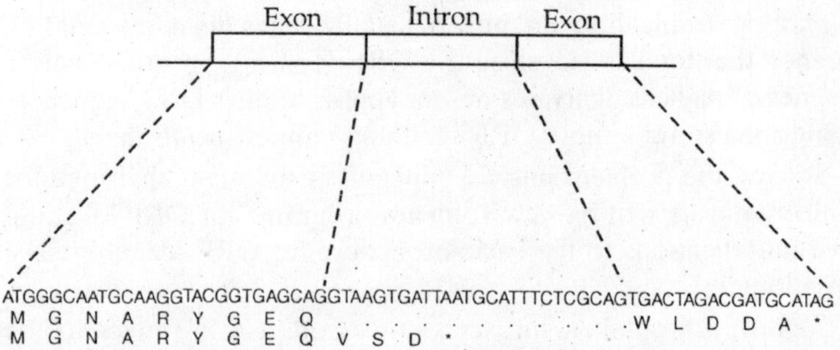

Figure 1.2. ORF scans are complicated by introns.

3. *Upstream control sequences* can be used to locate the regions where genes begin. This is because these control sequences, like exon-intron boundaries, have distinctive sequence features that they possess in order to carry out their role as recognition signals for the DNA-binding proteins involved in gene expression. Unfortunately, as with exon-intron boundaries, the control sequences are variable, more so in eukaryotes than prokaryotes, and in eukaryotes not all genes have the same collection of control sequences. Using these to locate genes is therefore problematical.

These three extensions of simple ORF scanning are generally applicable to all higher eukaryotic genomes. Additional strategies are also possible with individual organisms, based on the special features of their genomes. For example, vertebrate genomes contain *CpG islands* upstream of many genes, these being sequences of approximately 1 kb in which the GC content is greater than the average for the genome as a whole. For example, some 56% of human genes are associated with an upstream CpG island. These sequences are distinctive and when one is located in vertebrate DNA a strong assumption can be made that a gene begins in the region immediately downstream.

1.1.1.3 Homology searches give an extra dimension to sequence inspection

The limitations of ORF scanning with higher eukaryotic genomes are offset to a certain extent by the use of a *homology search* to test whether a series of triplets is a real exon or a chance sequence. In this analysis the DNA databases are searched to determine if the test sequence is identical or similar to any genes that have already been sequenced. Obviously, if the test sequence is part of a gene that has already been sequenced by someone else then an identical match will be found, but this is not the point of a homology search. Instead the intention is to determine if an entirely new sequence is similar to any known genes, because if it is then there is a chance that the test and match sequences are *homologous*, meaning that they represent genes that are evolutionarily related. The main use of homology searching is to assign functions to newly discovered genes, and we will therefore return to it when we deal with this aspect of genome analysis later in the chapter. At this point, we will note simply that the technique is also central to *gene location,* as it enables tentative exon sequences located by ORF scanning to be tested for functionality. If the tentative exon sequence gives one or more positive matches after a homology search then it probably is a real exon, if it gives no match then its authenticity must remain in doubt until it is assessed by one or other of the experiment-based gene location techniques. The value of this approach is increasing rapidly as the numbers of authentic gene sequences in databases multiply.

1.1.2 Experimental Techniques for Gene Location

It is to the experiment-based gene location techniques that we now turn our attention. These procedures locate genes by examining the RNA molecules that are transcribed from a DNA fragment. All genes are transcribed into RNA, and if the gene is discontinuous then the primary transcript is subsequently processed to remove the introns and link up the exons. Techniques which map the positions of transcribed (or 'expressed') sequences in a DNA fragment can therefore be used to locate exons and entire genes. The only problem to be kept in mind is that the transcript is usually longer than the coding part of the gene because it begins several tens of nucleotides upstream of the initiation codon and continues several tens or hundreds of nucleotides downstream of the termination codon. Transcript analysis does not therefore give a precise definition of the start and end of the coding region of a gene, but it does tell you that a gene is present in a particular region and it can locate the exon-intron boundaries. Often this is sufficient information to enable the coding region to be delineated.

1.1.2.1 Hybridization tests can determine if a fragment contains expressed sequences

The simplest procedures for studying expressed sequences are based on hybridization analysis. RNA molecules can be separated by specialized forms of agarose gel electrophoresis and transferred to a nitrocellulose or nylon membrane by the process called *northern blotting*. This differs from Southern blotting only in the precise conditions under which the transfer is carried out and the fact that it was not invented by a Dr Northern and so does not have a capital 'N'. If a northern blot is probed with a labeled DNA fragment, then RNAs expressed from the fragment will be detected. Northern hybridization is therefore, theoretically, a means of determining the number of genes present in a DNA fragment and the size of each coding region. There are two weaknesses with this approach

1. Some individual genes give rise to two or more transcripts, of different lengths, because some of their exons are optional and may or may not be retained in the mature RNA. If this is the case, then a fragment that contains just one gene could detect two or more hybridizing bands in the northern blot. A similar problem can occur if the gene is a member of a multigene family.
2. With many species, it is not practical to make a preparation from an entire organism so the RNA extract is obtained from a single organ or tissue. Consequently any genes not expressed in that organ or tissue will not be represented in the RNA population, and so will not be detected when the RNA is probed with the DNA fragment being studied. Even if the whole organism is used, not all genes will give hybridization signals because many are expressed only at a particular developmental stage, and others are weakly expressed, meaning that their RNA products are present in amounts too low to be detected by hybridization analysis.

A second type of hybridization analysis avoids the problems with poorly expressed and tissue-specific genes by searching not for RNAs but for related sequences in the DNAs of other organisms. This approach, like homology searching, is based on the fact that homologous genes in related organisms have similar sequences, whereas the noncoding DNA is quite different. If a DNA fragment from one species is used to probe a Southern blot of DNAs from related species, and one or more hybridization signals are obtained, then it is likely that the probe contains one or more genes. This is called *zooblotting*.

1.1.2.2 cDNA sequencing enables genes to be mapped within DNA fragments

Northern hybridization and zoo-blotting enable the presence or absence of genes in a DNA fragment to be determined, but give no positional information relating to the location of those genes in the DNA sequence. The easiest way to obtain this information is to sequence the relevant cDNAs. A cDNA is a copy of an mRNA and so corresponds to the coding region of a gene, plus any leader or trailer sequences' that are also transcribed. Comparing a cDNA sequence with a genomic DNA sequence therefore delineates the position of the relevant gene and reveals the exon-intron boundaries.

The degree of success of cDNA sequencing as a means of gene location depends on two factors. The first is the frequency of the

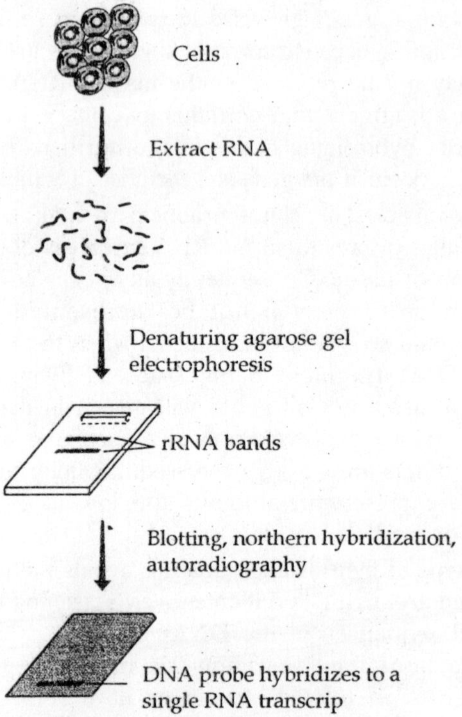

Figure 1.3. Northern hybridization.

appropriate cDNAs in the cDNA clone library that has been prepared. As with northern hybridization, the problem relates to the different expression levels of different genes. If the DNA fragment being studied contains one or more poorly expressed genes, then the relevant cDNAs will be rare in the library and it might be necessary to screen many clones before the desired one is identified. To get around this problem various methods of *cDNA capture* or *cDNA selection* have been devised, based around repeated hybridization between the DNA fragment being studied and the pool of cDNAs. Because the cDNA pool contains so many different sequences it is generally not possible to discard all irrelevant clones by these repeated hybridizations, but it is possible to enrich the pool for those clones that specifically hybridize to the DNA fragment. This reduces the size of the library that must subsequently be screened under stringent conditions to identify the desired clones.

A second factor determining success or failure is the completeness of the individual cDNA molecules. Traditionally, cDNAs are made by copying RNA molecules into single-stranded DNA with *reverse transcriptase* and then converting the single-stranded DNA into double stranded DNA with a DNA polymerase. There is always a chance that one or other of the strand synthesis reactions will not proceed to completion, resulting in a truncated cDNA. The presence of intramolecular base pairs in the RNA can also lead to incomplete copying. Truncated cDNAs may lack some of the information needed to locate the start and end points of a gene and all its exon-intron boundaries.

1.1.2.3 Methods are available for precise mapping of the ends of transcripts

The problems with incomplete cDNAs means that more robust methods are needed for locating the precise start and end points of gene transcripts. One possibility is a special type of PCR which uses RNA rather than DNA as the starting material. The first step in this type of PCR is to convert the RNA into cDNA with reverse transcriptase, after which the cDNA is amplified with *Taq* polymerase in the same way as in a normal PCR. These methods go under the collective name of *RT-PCR* (*reverse transcriptase PCR*) but the particular version that interests us at present is *RACE* (*rapid amplification of cDNA ends*). In the simplest form of this method one of the primers is specific for an internal region close to the beginning of the gene being studied. This primer attaches to the mRNA for the gene and directs the first, reverse transcriptase catalyzed, stage of the process, during which a cDNA corresponding to the start of the mRNA is made. Because only a small segment of the mRNA is being copied the expectation is that the cDNA synthesis will not terminate prematurely, so one end of the cDNA will correspond exactly with the start of the mRNA. Once the cDNA has been made a short linker is attached to its 3' end. The second primer anneals to this linker and, during the first round of the normal PCR, converts the single-stranded cDNA into a double-stranded molecule, which is subsequently amplified as the PCR proceeds. The sequence of this amplified molecule will reveal the precise position of the start of the transcript.

Other methods for precise transcript mapping involve *heteroduplex analysis*. If the DNA region being studied is cloned as a restriction fragment in an M13 vector then it can be obtained as single-stranded DNA. When mixed with an appropriate RNA preparation the transcribed sequence in the cloned DNA hybridizes with the equivalent mRNA, forming a double-stranded heteroduplex. In the example, the start of this mRNA lies within the cloned restriction fragment, so some of the cloned fragment participates in the heteroduplex, but the rest does not. The single-stranded regions can be digested by treatment with a single-strand-specific nuclease such as S_1. The size of the heteroduplex is determined by degrading the RNA component with alkali and electrophoresing the single-stranded DNA in an agarose gel. This size measurement is then used to position the start of the transcript relative to the restriction site at the end of the cloned fragment.

1.1.2.4 Exon-intron boundaries can also be located with precision

Heteroduplex analysis can also be used to locate exon-intron boundaries. The method is exactly the same, with the exception that the cloned restriction fragment spans not the start of the transcript but the exon-intron boundary being mapped.

A second method for finding exons in a genome sequence is called *exon trapping*. This requires a special type of vector that contains a *minigene* consisting of two exons flanking an intron sequence, the first exon being preceded by the sequence signals needed to initiate transcription in a eukaryotic cell. To use the vector the

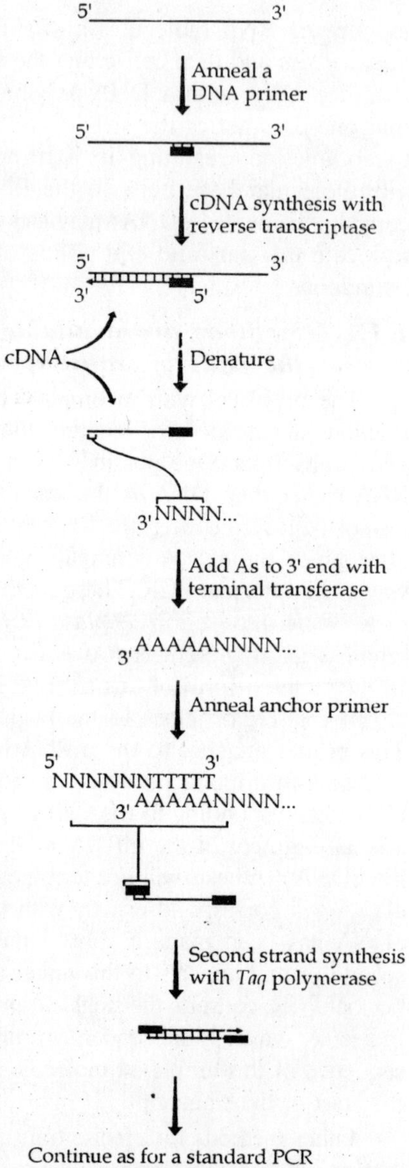

Figure 1.4. RACE–rapid amplification of cDNA ends.

piece of DNA to be studied is inserted into a restriction site located within the vector's intron region. The vector is then introduced into a suitable eukaryotic cell line, where it is transcribed and the RNA produced from it is spliced. The result is that any exon contained in the genomic fragment becomes attached between the upstream and downstream exons from the minigene. RT-PCR with primers annealing within the two minigene exons is now used to amplify a DNA fragment which is sequenced. As the minigene sequence is already known the nucleotide positions at which the inserted exon starts and ends can be determined, precisely delineating this exon.

1.1.3 Determining the Function of a Gene

Once a new gene has been located, the question of its function has to be addressed. This is turning out to be a difficult area of genomics research, completed sequencing projects revealing that we know rather less than we thought about the content of individual genomes. *E. coli* and *S. cerevisiae*, for example, were intensively studied by conventional genetic analysis prior to the advent of sequencing projects, and geneticists were at one time fairly confident that most of their genes had been identified by mutation studies. The genome sequences revealed that in fact there are large gaps in our knowledge. Of the 4288 protein-coding genes in the *E. coli* genome sequence, only 1853 (43% of the total) had been previously identified. For *S. cerevisiae* the figure was only 30%. As with gene location, attempts to determine the functions of unknown genes are made by computer analysis and by experimental studies.

1.1.4 Computer Analysis of Gene Function

We have already seen that computer analysis plays an important role in locating genes in DNA sequences, and that one of the most powerful tools available for this purpose is homology searching, which locates genes by comparing the DNA sequence under study with all the other DNA sequences in the databases. The basis of homology searching is that related genes have similar sequences and so a new gene can be discovered by virtue of its similarity to an equivalent, already sequenced, gene from a different organism. Now we will look more closely at homology analysis and see how it can be used to assign a function to a new gene.

1.1.5 Homology Reflects Evolutionary Relationships

Homologous genes are ones that share a common evolutionary ancestor, revealed by sequence similarities between the genes. These similarities form the data on which molecular phylogenies are based. Homologous genes fall into two categories:

1. *Orthologous* genes are those homologs that are present in different organisms and whose common ancestor predates the split between the species.
2. *Paralogous* genes are present in the same organism, often members of a recognized multigene family their common ancestor possibly or possibly not predating the species in which the genes are now found.

A pair of homologous genes do not usually have identical nucleotide sequences, because the two genes undergo different random changes by mutation, but they have similar sequences because these random changes have operated on the same starting sequence, the common ancestral gene. Homology searching makes use of these sequence similarities. The basis of the analysis is that if a newly sequenced gene turns out to be similar to a second, previously sequenced gene, then an evolutionary relation-ship can be inferred and the function of the new gene is likely to be the same, or at least similar, to the function of the second gene.

It is important not to confuse the words *homology* and *similarity*. It is incorrect to describe a pair of related genes as '80% homologous' if their sequences have 80% nucleotide identity. A pair of genes are either evolutionarily related or they are not, there are no in between situations and it is therefore meaningless to ascribe a percentage value to homology.

1.1.5.1 Homology analysis can provide information on the function of an entire gene or of segments within it

A homology search can be conducted with a DNA sequence but usually a tentative gene sequence is converted into an amino acid sequence before the search is carried out. One reason for this is because there are twenty different amino acids in proteins but only four nucleotides in DNA, so genes that are unrelated usually appear to be more different from one another when their amino acid sequences are compared. A homology search is therefore less likely to give spurious results if the amino sequence is used. The practicalities of homology searching are not at all daunting. Several software programs exist for carrying out this type of analysis, the most popular being BLAST. The search can be carried out simply by sending a correctly formatted email to the BLAST server at a

DNA database. Several hours later the server sends you back an email giving the results of the search. There is currently approximately a 50% chance that a newly sequenced gene will-give a positive homology match with a gene already in the databases.

A positive match may give a clear indication of the function of the new gene or the implications of the match might be more subtle. In particular, genes that have no obvious evolutionary relatedness might have short segments that are similar to one another. The explanation of this is often that, although the genes are unrelated, their proteins have similar functions and the shared sequence encodes a domain within each protein that is central to that shared function. Although the genes themselves have no common ancestor, the domains do, but with their common ancestor occurring at a very ancient time, the homologous domains having subsequently evolved not only by single nucleotide changes, but also by more complex rearrangements that have created new genes within which the domains are found. An interesting example is provided by the tudor domain, an approximately 120-amino-acid motif which was first identified in the sequence of the *Drosophila melanogaster* gene called *tudor*. The protein coded by the *tudor* gene, whose function is unknown, is made up often copies of the tudor domain, one after the other. A homology search using the tudor domain as the test revealed that several known proteins contain this domain. The sequences of these proteins are not highly similar to one another and there is no indication that they are true homo logs, but all possess the tudor domain. These proteins include one involved in RNA transport during *Drosophila* oogenesis, a human protein with a role in RNA metabolism, and others whose activity appears to involve RNA in one way or another. The homology analysis therefore suggests that the tudor sequence plays some part in RNA binding,

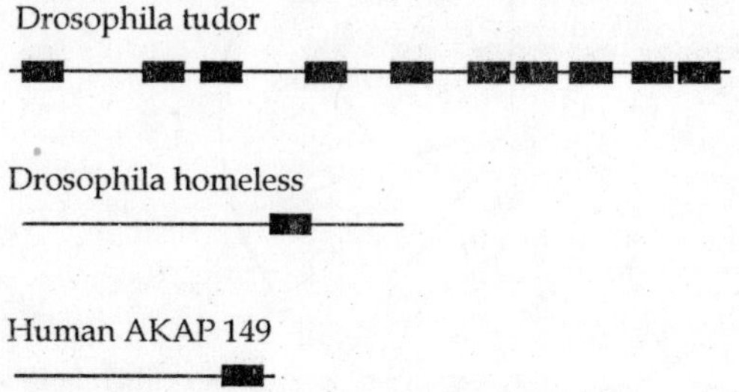

Figure 1.5. The tudor domain. The top drawing shows the structure of the *Drosophila tudor* protein, which contains ten copies of the tudor domain.

RNA metabolism or some other function involving RNA. The information from the computer analysis is incomplete by itself, but it points the way to the types of experiment that should be done to obtain more clear-cut data on the function of the tudor domain.

1.1.6 Homology Analysis in the Yeast Genome Project

The S. *cerevisiae* genome project has illustrated both the potential and limitations of homology analysis as a means of assigning functions to new genes. The yeast genome contains approximately 6000 genes, of which 30% had been identified by conventional genetic analysis before the sequencing project got underway. The remaining 70% were studied by homology analysis, giving the following results.

1. Almost another 30% of the genes in the genome could be assigned functions after homology searching of the sequence databases. About half of these were clear homologs of genes whose functions had previously been established, and about half had less striking similarities, including many where the similarities were restricted to discrete domains. For all these genes the homology analysis could be described as successful, but with various degrees of usefulness. For some genes the identification of a homolog enabled the function of the yeast gene to be comprehensively determined: examples included identification of yeast genes for DNA polymerase subunits and aminoacyl-tRNA synthetases. For other genes the functional assignment could only be to a broad category, such as 'gene for a protein kinase', in other words the biochemical properties of the gene product

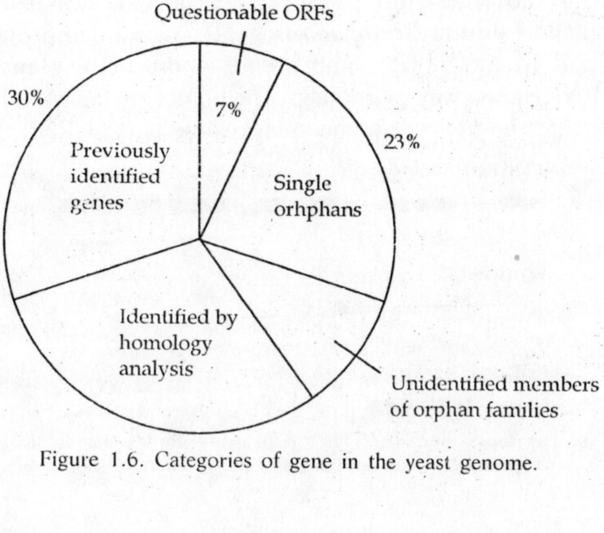

Figure 1.6. Categories of gene in the yeast genome.

could be inferred but not the exact role of the protein in the cell. Some identifications were initially puzzling, the best example being the discovery of a yeast homolog of a bacterial gene involved in nitrogen fixation. Yeasts do not fix nitrogen so this could not be the function of the yeast gene. In this case, the discovery of the yeast homolog refocused attention on the previously characterized bacterial gene, with the subsequent realization that, although being involved in nitrogen fixation, the primary role of the bacterial gene product was in the synthesis of metal-containing proteins, which have broad roles in all organisms, not just nitrogen-fixing ones.

2. About 10% of all the yeast genes had homologs in the databases, but the functions of these homologs were unknown. The homology analysis was therefore unable to help in assigning functions to these yeast genes. These yeast genes and their homologs are called *orphan families*.
3. The remaining yeast genes, about 30% of the total, had no homologs in the databases. A small proportion of these (about 7% of the total) were questionable ORFs which might not be real genes, being rather short or having an unusual codon bias. The remainder look like genes but at present are unique. The latter group are called single orphans.

1.1.7 Assigning Gene Function by Experimental Analysis

It is clear that homology analysis is not a panacea that can identify the functions of all new genes. Experimental methods are therefore needed to complement and extend the results of homology studies. This is proving to be one of the biggest problems in genomics research and most molecular biologists agree that the methodologies and strategies currently in use are not entirely adequate for assigning functions to the vast numbers of unknown genes being discovered by sequencing projects. The problem is that the objective – to plot a course from gene to function – is the reverse of the route normally taken by genetic analysis, in which the starting point is a phenotype and the objective is to identify the underlying gene or genes. The problem we are currently addressing takes us in the opposite direction, starting from a new gene and hopefully leading to identification of the associated phenotype.

1.1.8 Gene Inactivation is the key to Functional Analysis

In conventional genetic analysis, the genetic basis of a phenotype is usually studied by searching for mutant organisms in which the phenotype has become altered. The mutants might be obtained experimentally, for example by treating a population of organisms (*e.g.* a culture of bacteria) with ultraviolet radiation or a mutagenic chemical or the mutants might be present in a natural population. The gene or genes that have been altered in the mutant organism are then studied by genetic crosses, which can locate the position of a gene in a genome and also determine if the gene is the same as one that has already been characterized. The gene can then be studied further by molecular biology techniques, for example by cloning and sequencing.

The general principle of this conventional analysis is that the genes responsible for a phenotype can be identified by determining which genes are inactivated in organisms that display a mutant version of the phenotype. If the starting point is the gene, rather than the phenotype, then the equivalent strategy would be to mutate the gene and identify the phenotypic change that results. This is the basis of most of the techniques used to assign functions to unknown genes.

1.1.8.1 Methods for inactivating specific genes depend on homologous recombination

The easiest way to inactivate a specific gene is to disrupt it with an unrelated segment of DNA. This can be achieved by *homologous recombination* between the chromosomal copy of the gene and a second piece of DNA that shares some sequence identity with the target gene. Homologous, and other types of, recombination are complex events, for present purposes it is enough to know that if two DNA molecules have similar sequences, then recombination can result in segments of the molecules being exchanged.

How is gene inactivation carried out in practice? We will consider two examples, the first with *S. cerevisiae*. Since completing the genome sequence in 1996, yeast molecular biologists have embarked on a coordinated, international effort to determine the functions of as many orphan genes as possible. The central component is the 'deletion cassette', which carries a gene for antibiotic resistance. This gene is not a normal component of the yeast genome but it will work if transferred into a yeast chromosome, giving rise to a transformed yeast cell that is resistant to the antibiotic geneticin. Before using the deletion cassette new segments of DNA are attached as tails to either end. These segments have sequences identical to parts of the yeast gene that is going to be inactivated. After the

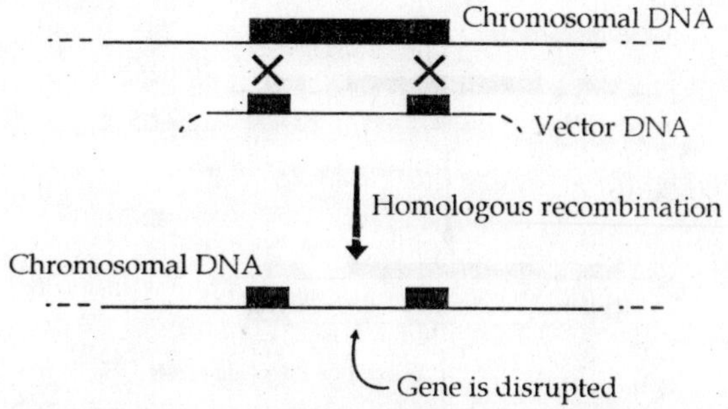

Figure 1.7. Gene inactivation by homologous recombination.

modified cassette is introduced into a yeast cell, homologous recombination occurs between the DNA tails and the chromosomal copy of the yeast gene, replacing the latter with the antibiotic resistant gene. Cells which have undergone the replacement are therefore selected by plating the culture on to agar medium containing geneticin. The resulting colonies lack the target gene and their phenotypes can be examined to gain some insight into the function of the gene.

The second example of gene inactivation uses an analogous process but with mice rather than yeast. The mouse is frequently used as a *model organism* for humans, because the mouse genome is similar to the human genome, containing many of the same genes. Identifying the functions of unknown human genes is therefore being carried out largely by inactivating the equivalent genes in the mouse, these experiments being ethically unthinkable with humans. The homologous recombination part of the procedure is identical to that described for yeast and once again results in a cell in which the target gene has been inactivated. The problem is that we do not want just one mutated cell, we want a whole mutant mouse, as only with the complete organism can we make a complete assessment of the effect of the gene inactivation on the phenotype. To achieve this it is necessary to use a special type of mouse cell, an *embryonic stem* or ES cell.

Unlike most mouse cells, ES cells are *totipotent*, meaning that they are not committed to a single developmental pathway and can therefore give rise to all types of differentiated cell. The engineered ES cell is therefore injected into a mouse embryo, which continues to develop and eventually give rise to a *chimera*, a mouse whose cells are a mixture of mutant ones, derived from the engineered ES cells, and nonmutant ones, derived from all the

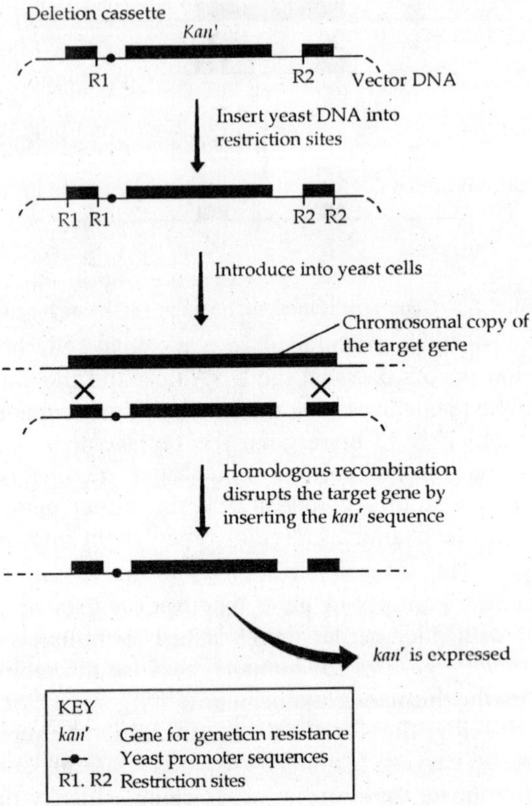

Figure 1.8. The use of a yeast deletion cassette.

other cells in the embryo. This is still not quite what we want, so the chimeric mice are allowed to mate with one another. Some of the offspring result from fusion of two engineered gametes, and will therefore be nonchimeric, as every one of their cells will carry the inactivated gene. These are *knockout mice,* and it is hoped that examination of their phenotypes provides the desired information on the function of the gene being studied. This works well for many gene inactivations but some are lethal and so cannot be studied in a homozygous knockout mouse. Instead, a heterozygous mouse is obtained, the product of fusion between one normal and one engineered gamete, in the hope that the phenotypic effect of the gene inactivation will be apparent even though the mouse still has one correct copy of the gene being studied.

1.1.8.2 Phenotypic effect of gene inactivation is sometimes difficult to discern

Once a gene-inactivated yeast strain, knockout mouse, or equivalent with any other organism has been obtained the next stage is to examine the phenotype of the mutant in order to assign a function to the unknown gene. This can be much more difficult than it sounds. With any organism the range of phenotypes that must be examined is immense. Even with yeast, the list is quite lengthy and with higher organisms some phenotypes (*e.g.* behavioural ones) are difficult if not impossible to assess in a comprehensive fashion. Furthermore, the effects of some gene inactivations are very subtle and may not be recognized when the phenotype is examined. A good example was provided by the longest gene on yeast chromosome III which, at 2167 codons and with typical yeast codon bias, simply had to be a functional gene rather than a spurious ORE. Inactivation of this gene had no apparent effect, the mutant yeast cells appearing to have an identical phenotype to normal yeast. For some time it was thought that perhaps this gene is dispensable, its protein product either involved in some completely nonessential function, or having a function that is duplicated by a second gene. Eventually it was shown that the mutants died when they were grown at low pH in the presence of glucose and acetic acid, which normal yeasts can tolerate, and it was concluded that the gene codes for a protein that pumps acetate out of the cell. This is definitely an essential function as the gene plays a vital role in protecting yeast from acetic acid induced damage, but this essentiality was difficult to track down from the phenotype tests.

These problems, together with the daunting prospect of having to perform a gene inactivation and full phenotype screen for everyone of the 2500 yeast orphans, has prompted yeast biologists to pioneer the search for more efficient methods of assigning functions to unknown genes.

1.1.9 Gene Overexpression can also be Used to Assess Function

So far we have concentrated on techniques that result in inactivation of the gene being studied ('loss of function'). The complementary approach is to engineer an organism in which the test gene is much more active than normal ('gain of function') and to determine what changes, if any, this has on the phenotype. The results of these experiments must be treated with caution because

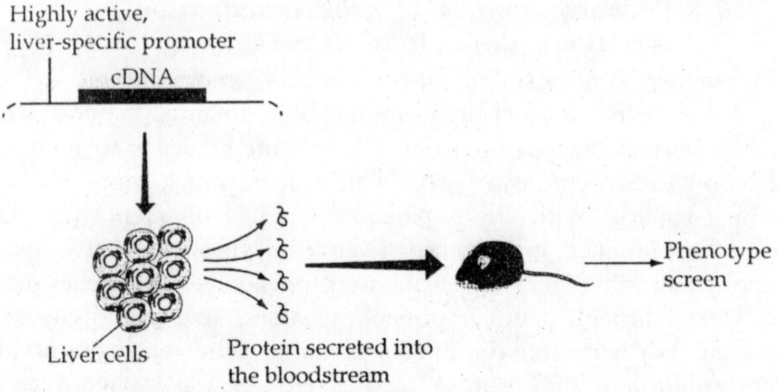

Figure 1.9. Functional analysis by gene overexpression.

of the need to distinguish between a phenotype change that is due to the specific function of an over expressed gene and a less specific phenotype change that reflects the abnormality of the situation where a single gene product is being synthesized in excessive amounts, possibly in tissues in which the gene is normally inactive. Despite this qualification, over expression has provided some important information on gene function.

To overexpress a gene a special type of cloning vector must be used, one designed to ensure that the cloned gene directs the synthesis of as much protein as possible. The vector is therefore *multicopy*, meaning that inside the host organism it multiplies to 40-200 copies per cell, so there are many copies of the test gene. The vector must also contain a highly active promoter so that each copy of the test gene is converted into large quantities of mRNA, again ensuring that as much protein as possible is made. In this project the genes to be studied were identified from EST sequences and so each one was cloned as a cDNA. The genes were selected because their sequences suggested that they code for proteins that are secreted into the bloodstream. The cloning vector that was used contained a highly active promoter that is expressed only in the liver, so each *transgenic mouse* overexpressed the test gene in its liver and then secreted the resulting protein into the blood.

The phenotype of each transgenic mouse was examined in the search for clues regarding the functions of the cloned genes. An interesting discovery was made when it was realized that one transgenic mouse had bones that were significantly more dense than those of normal mice. This was important for two reasons: first, it enabled the relevant gene to be identified as one involved in bone synthesis; and second, the discovery of a protein that increases bone density has implications for the development of treatments for human osteoporosis, a fragile-bone disease.

1.1.9.1 More detailed studies of the activity of a protein coded by an unknown gene

Gene inactivation and overexpression are the primary techniques used by genome researchers to determine the function of a new gene, but these are not the only procedures that can provide information on gene activity. Other methods can extend and elaborate the results of inactivation and overexpression. These can be used to provide additional information that will aid identification of a gene function, or might form the basis of a more comprehensive examination of the activity of a protein whose gene has already been characterized.

1.1.9.2 Directed mutagenesis can be used to probe gene function in detail

Inactivation and overexpression can determine the overall function of a gene, but they cannot provide detailed information on the activity of a protein coded by a gene. For example, it might be suspected that part of a gene codes for an amino acid sequence that directs its protein product to a particular compartment in the cell, or is responsible for the ability of the protein to respond to a chemical or physical signal. To test these hypotheses it would be necessary to delete or alter the relevant part of the gene sequence, but to leave the bulk unmodified so that the protein is still synthesized and retains the major part of its activity. The various procedures of *site-directed* or *in vitro* mutagenesis can be used to make these subtle changes. These are important techniques whose applications lie not only with the study of gene activity but also in the area of *protein engineering*, where the objective is to create novel proteins with properties that are better suited for use in industrial or clinical settings.

After mutagenesis the gene sequence must be introduced into the host cell so that homologous recombination can replace the existing copy of the gene with the modified version. This presents a problem because we must have a way of knowing which cells have undergone homologous recombination. Even with yeast this will only be a fraction of the total, and with mice the fraction will be very small. Normally we would solve this problem by placing a marker

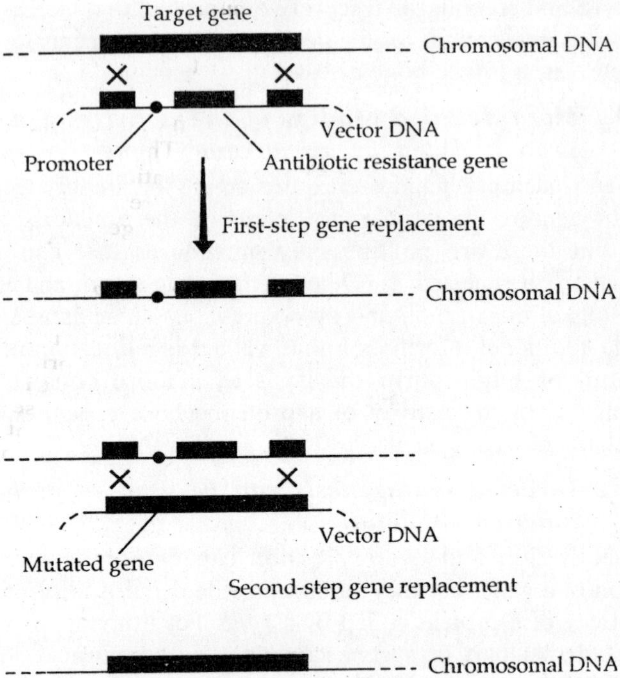

Figure 1.10. Two-step gene replacement.

gene (*e.g.* one coding for antibiotic resistance) next to the mutated gene and looking for cells that take on the phenotype conferred by this marker. In most cases, cells that insert the marker gene into their genome also insert the closely attached, mutated gene and so are the ones we want. The problem is that in a site-directed mutagenesis experiment we must be sure that any change in the activity of the gene being studied is due to the specific mutation that was introduced into the gene rather than to the indirect result of changing its environment in the genome by inserting a marker gene next to it. The answer is to use a more complex, two-step gene replacement. In this procedure the target gene is first replaced with the marker gene, on its own, the cells in which this recombination takes place being identified by selecting for the marker gene phenotype. These cells are then used in the second stage of the gene replacement, when the marker gene is replaced by the mutated gene, success now being monitored by looking for cells that have lost the marker gene phenotype. These cells contain the mutated gene and their phenotypes can be examined to determine the effect of the directed mutation on the activity of the protein product.

1.1.9.3 Reporter genes and immunocytochemistry can be used to locate where and when genes are expressed

Clues to the function of a gene can often be obtained by determining where and when the gene is expressed. If gene expression is restricted to a particular organ or tissue of a multicellular organism, or to a single set of cells within an organ or tissue, then this positional information can be used to infer the general role of the gene product. The same is true of information relating to the developmental stage at which a gene is expressed, this type of analysis having proved particularly useful in understanding the activities of genes involved in the earliest stages of *Drosophila* development and increasingly being used to unravel the genetics of mammalian development. It is also applicable to those unicellular organisms, such as yeast, which have distinctive developmental stages in their life cycle.

Determining the pattern of gene expression within an organism is possible with a *reporter gene*. This is a gene whose expression can be monitored in a convenient way, ideally by visual examination, cells that express the reporter gene becoming blue, fluorescing or giving off some other visible signal. For the reporter gene to give a reliable indication of where and when a test gene is expressed the reporter must be subject to the same regulatory signals as the test gene. This is achieved by replacing the ORF of the test gene with the ORF of the reporter gene, the regulatory signals being contained in the region of DNA upstream of an ORF. The reporter gene will now display the same expression pattern as the test gene, this expression pattern being determined by examining the organism for the reporter signal.

As well as knowing in which cells a gene is expressed, it is often useful to locate the position within the cell where the protein coded by the gene is found. For example, key data regarding gene function can be obtained by showing that the protein product is located in mitochondria, in the nucleus, or on the cell surface. Reporter genes cannot help here as the DNA sequence upstream of the gene, the sequence to which the reporter gene is attached, is not involved in targeting the protein product to its correct intracellular location. Instead it is the amino acid sequence of the protein itself that is important. Therefore the only way to determine where the protein is located is to search directly for it. This is done by *immunocytochemistry*, which makes use of an antibody that is specific for the protein of interest and so binds to this protein and no other. The antibody is labeled so that its position in the cell, and hence the position of the target protein, can be visualized. For low resolution studies, fluorescent labeling and light microscopy are used; alternatively, high resolution immunocytochemistry can be carried out by electron microscopy with an electron dense label such as colloidal gold.

1.1.10 Protein-Protein Interactions can be Examined by Phage Display and the Yeast Two-hybrid System

Important data pertaining to gene function can often be obtained by determining if the protein coded by the gene interacts with other, known proteins. If an interaction to a well characterized protein is identified, then it might be possible to infer the function of the unknown protein. For example, an interaction between an unknown protein and one known to be located on the cell surface might indicate that the unknown protein is involved in cell-cell signaling. There are several methods for studying protein-protein interactions, the two most useful being *phage display* and the *yeast two-hybrid system*.

In phage display a special type of cloning vector is used. This vector is based on λ bacteriophage or one of the filamentous bacteriophages such as M13. It is designed so that a new gene that is cloned into it is expressed in such a way that its protein product is synthesized as a fusion with one of the phage coat proteins. The phage protein therefore carries the foreign protein into the phage coat, where it is 'displayed' in a form that enables it to interact with other proteins that the phage encounters. There are several ways in which phage display can be used to study protein interactions. In one method, the test protein is displayed and interactions sought with a series of purified proteins or protein fragments of known function. This approach is limited as it takes time to carry out each test, so is feasible only if some prior information has been obtained about likely interactions. A more powerful strategy is to prepare a phage display library, a collection of clones displaying a range of proteins, and identifying which members of the library interact with the test protein.

The yeast two-hybrid system detects protein interactions in a more complex way. The proteins called *transcription factors* are responsible for controlling the expression of genes in eukaryotes. To carry out this function a transcription factor must bind to a DNA sequence upstream of a gene and activate the RNA polymerase enzyme that copies the gene into RNA. These two activities are specified by different parts of the transcription factor's protein structure, and some factors will work even after the protein has been cleaved into two segments, one segment containing the DNA-binding domain and one the activation domain. In the cell, the two segments interact to form the functional transcription factor.

The two-hybrid system makes use of an *S. cerevisiae* strain that lacks a transcription factor for a reporter gene. This gene is therefore switched off. The two segments of the transcription factor are coded by separate genes, both of which have been cloned. One of these genes, usually the one for the DNA-binding domain, is ligated to the gene for the protein whose interactions we wish to study. This protein can come from any organism, not just yeast: in the example it is a human protein, which for convenience we will call 'Protein X'.

After introduction into yeast, this engineered gene will specify synthesis of a fusion protein, made up of the DNA-binding domain of the transcription factor attached to Protein X. The second gene, the one for the activation domain of the transcription factor, is ligated with a mixture of DNA fragments so that many different constructs are made. The engineered genes are mixed together with yeast cells, some of which become cotransformed with an engineered DNA-binding gene and an engineered activation gene. If the fusion proteins coded by these two genes can cooperate then the reporter gene will be switched on. If this happens then the protein attached to the activation domain of the transcription factor must be one that can interact with Protein X.

1.2 COMPARATIVE GENOMICS

We have already seen how similarities between homologous genes from different organisms provide one way of assigning a function to an unknown gene. This is an example of how knowledge about the genome of one organism can help in understanding the genome of a second organism. The possibility that comparative genomics might be a valuable means of deciphering the human genome was recognized when the Human Genome Project was planned in the late 1980s, and the Project has actively stimulated the development of genome projects for model organisms such as the mouse and fruit fly. In this section we will explore the extent to which comparisons between different genomes are proving useful.

1.2.1 Comparative Genomics as an Aid to Gene Mapping

The basis of comparative genomics is that the genomes of related organisms are similar. The argument is the same one that we considered when looking at homologous genes. Two organisms with a relatively recent common ancestor will have genomes that display species-specific differences built onto the common plan possessed by the ancestral genome. The closer two organisms are on the evolutionary scale, the more related their genomes will be.

If the two organisms are sufficiently closely related then their genomes might display synteny, the partial or complete conservation of gene order. Then it is possible to use map information from one genome to locate genes in the second genome. At one time it was thought that mapping the genomes of mouse and other mammals, which are at least partially syntenic with the human

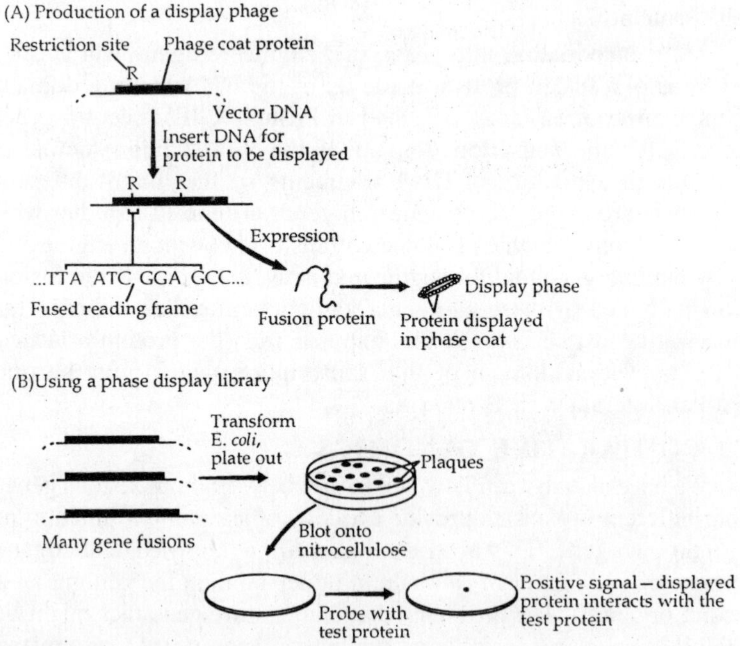

Figure 1.11. Phage display. (A) The cloning vector used for phage display is a bacteriophage genome with a unique restriction site located within a gene for a coat protein. (B) A phage display library is produced by ligating many different DNA fragments into the cloning vector.

genome, might provide valuable information that could be used in construction of the human genome map. The problem with this approach is that all the close relatives of humans have equally large genomes that are just as difficult to study, the only advantage being that a genetic map is easier to construct with an animal which, unlike humans, can be subjected to experimental breeding programs. Despite the limitations of human pedigree analysis, progress has been more rapid in mapping the human genome than those of any of our close relatives, so in this respect comparative genomics is proving more useful in mapping the animal genomes rather than our own. This is in itself a useful corollary to the Human Genome Project because it is revealing animal homologs of human genes involved in diseases, providing animal models in which these diseases can be studied.

Mapping is significantly easier with a small genome compared with a large one. This means that if one member of a pair of syntenic genomes is substantially smaller than the other, then mapping studies with this small genome is likely to provide a real boost to equivalent work with the larger genome. The pufferfish, *Fugu rubripes,* has been proposed in this capacity with respect to the human genome. The pufferfish genome is just 400 Mb, less than one-seventh the size of the human genome but containing approximately the same number of genes. The mapping work carried out to date with the pufferfish indicates that there is some similarity with the human gene order, at least over short distances. This means that it should be possible, to a certain extent, to use the pufferfish map position to find human homologs of sequenced pufferfish genes, and *vice versa*. This may be useful in locating unmapped human genes, but holds greatest promise in locating essential sequences such as promoters and other regulatory signals upstream of human genes. This is because these signals are likely to be similar in the two genomes, and recognizable because they are surrounded with noncoding DNA that has diverged quite considerably by random mutations.

One area where comparative genomics has a definite advantage is in the mapping of plant genomes. Wheat provides a good example. Wheat is the most important food plant in the human diet, being responsible for approximately 20% of the human calorific intake, and is therefore one of the crop plants that we most wish to study and possibly manipulate in the quest for improved crops.

Unfortunately, the wheat genome is huge at 17000Mb, more than five times larger than even the human genome. A small model genome with a gene order similar to that of wheat would therefore be useful as a means of mapping desirable genes which might then be obtained from their equivalent positions in the wheat genome. Wheat, and other cereals such as rice, are members of the *Gramineae*, a large and diverse family of grasses. The rice genome is only 400Mb, substantially smaller than that of wheat, and there are probably other grasses with even smaller genomes. Comparative mapping of the rice and wheat genomes has revealed many similarities, and the possibility therefore exists that genes from the wheat genome might be isolated by first mapping the positions of the equivalent genes in a smaller *Gramineae* genome.

1.2.2 Comparative Genomics in the Study of Human Disease Genes

One of the main reasons for sequencing the human genome is to gain access to the sequences of genes involved in human disease. The hope is that the sequence of a disease gene will provide an insight into the biochemical basis of the disease and hence indicate a way of preventing or treating the disease. Comparative genomics has an important role to play in the study of disease genes because the discovery of a homolog of a human disease gene in a second organism is often the key to understanding the biochemical function of the human gene. If the homolog has already been characterized then the information needed to understand the biochemical role of the human gene may already be in place; if it is not then the necessary research can be directed at the homolog.

To be useful in the study of disease-causing genes, the second genome does not need to be syntenic with the human genome, nor even particularly closely related. *Drosophila* holds great promise in this respect, as the phenotypic effects of many *Drosophila* genes are already well known, so the data already exist for inferring the mode of action of human disease genes that have homologs in the *Drosophila* genome. But the greatest success has been with yeast. Several human disease genes have homologs in the s. *cerevisiae* genome. These disease genes include ones involved in cancer, cystic fibrosis and neurological syndromes, and in several cases the yeast homolog has a known function that provides a clear indication of the biochemical activity of the human gene. In some cases it has even been possible to demonstrate a physiological similarity between the gene activity in humans and yeast. For example, the yeast gene *SGSI* is a homolog of a human gene involved in the premature aging disease called Werner's syndrome. Yeasts with a mutant *SGSI* gene live for shorter periods than normal yeasts and display accelerated onset of aging indicators such as sterility.

In yeast the underlying cause of the premature aging, and possibly the direct effect of the mutation, appears to be a change in the distribution of proteins in the nucleolus, a subcompartment of the nucleus. The link between *SGSI* and the Werner's syndrome gene, provided by comparative genomics, means that these findings are directly relevant to study of the human disease.

1.3 FROM GENOME TO CELL

Even if every gene in a genome can be identified and assigned a function, a challenge still remains. This is to understand how the genome as a whole operates within the cell, specifying and coordinating the various biochemical activities that take place. Describing and explaining this aspect of genome biology will be a mammoth task that will occupy researchers for many decades of the 21st century. Attempts are currently being made to chart the pattern of gene expression - which genes are switched on and which are switched off - in different tissues, during different developmental stages and, in the case of humans, in different disease states.

1.3.1 Patterns of Gene Expression

All organisms regulate expression of their genes, so that those genes whose protein products are not needed at a particular time are switched off. Understanding which genes are active in particular tissues at particular times, and assessing the relative degrees of activity of the genes that are switched on, is the first step in understanding how the genome as a whole functions within the cell.

1.3.2 Gene Expression is Assayed by Measuring Levels of RNA Transcripts

The first stage of gene expression involves copying the gene into an RNA transcript. Identifying if a transcript of a gene is present in a cell or tissue is therefore the most direct way of determining whether the gene is switched on. This can be done by immobilizing the DNA copy of the gene on a solid support and hybridizing it with a sample of RNA extracted from the cells being studied. In fact the RNA is usually converted into cDNA prior to being used as the hybridization probe, as cDNA synthesis is the most convenient

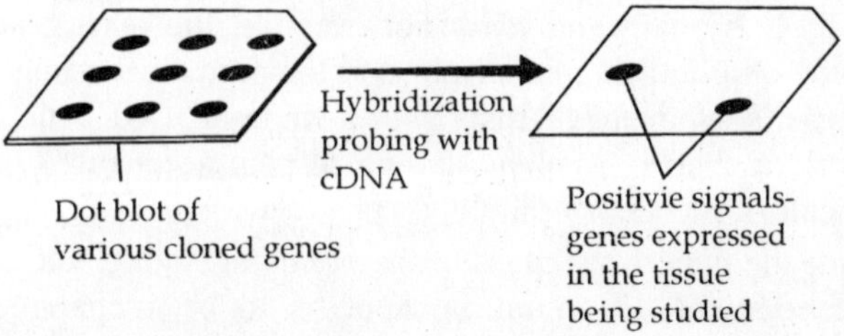

Figure 1.12. Assaying gene expression by dot blot hybridization analysis.

way of obtaining labeled sequences representing RNA molecules. If transcripts of the gene are present in the RNA extract then a hybridization signal will be seen.

Determining if a gene is being expressed is therefore a routine task. The complications become apparent when one appreciates that, in order to assay the entire mRNA content of a cell (referred to as its *transcriptome*), this basic procedure has to be repeated for every proteincoding gene in the genome. The answer is to use a DNA chip or other kind of microarray.

In an early experiment, all 6000 *S. cerevisiae* genes were immobilized as an 80 × 80 array in a 18 mm by 18 mm square on a glass slide, and changes in gene expression that accompany the transition from aerobic to anaerobic respiration assayed, using fluorescently labeled cDNA with hybridization of the cDNA to the microarray being visualized by scanning confocal microscopy. Twofold or greater changes in expression level, either up or down, were recorded for 1740 genes, over a quarter of the total, including many orphans whose functions are not known.

Not surprisingly, applying these and equivalent techniques to studies of human gene expression patterns has proved more difficult. Most effort has been directed at cancerous tissues because understanding the abnormal features of gene expression in these cells might lead to a means of diagnosing the disease state and monitoring progress during treatment, and might even point the way towards new therapies.

One study discovered 289 genes whose expression patterns differed significantly when normal colon epithelial cells were compared with colon cancer cells, about half of these genes also showing abnormal expression levels in pancreatic cancer cells. The implication is that some genes are abnormally expressed in a range of cancers and some are specific for a particular type of cancerous tissue.

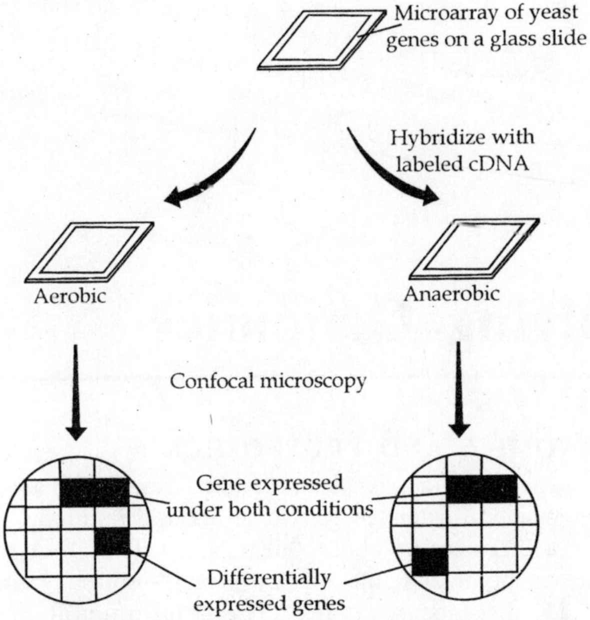

Figure 1.13. Assaying gene expression with a microarray.

1.3.3 Proteome

In the second stage of gene expression, mRNA transcripts are translated into protein, the entire protein content of the cell being referred to as the *proteome*. Determining the identities and relative amounts of the proteins in a cell or tissue is more difficult than transcript studies because techniques for protein characterization are less well advanced than the equivalent RNA procedures. Delineating the proteome is, however, one of the keys to understanding how the genome as a whole is expressed. Techniques at present are based around two-dimensional polyacrylamide gel electrophoresis to separate proteins, or peptides derived from proteins by proteolytic cleavage, followed by mass spectrometry to determine the amino acid compositions of the protein or peptide in each spot on the two-dimensional gel. This information can be related to the sequences of all the proteincoding genes in the genome in order to identify which gene specifies which spot. A change in the proteome due, for example, to cellular differentiation, is revealed by a change in the pattern of spots.

2

Evolving Genomics

2.1 GENOMICS AND PROTEOMICS

The genome of a typical bacterium comes as a single DNA molecule, which, if extended, would be about 2 mm long. (The cell itself has a diameter of about 0.001 mm.) The DNA of higher organisms is organized into chromosomes—normal human cells contain 23 chromosome pairs. The total amount of genetic information per cell—the sequence of nucleotides of DNA—is very nearly constant for all members of a species, but varies widely between species:

Organism	*Genome size*	
Epstein-Barr virus	0.172×10^6	base pairs
Bacterium (*E. coli*)	4.6×10^6	
Yeast (*S. cerevisiae*)	12.1×10^6	
Nematode worm (*C. elegans*)	95.5×10^6	
Thale cress (*A. thaliana*)	117.0×10^6	
Fruit fly (*D. melanogaster*)	180.0×10^6	
Human (*H. sapiens*)	3200×10^6	

Not all DNA codes for proteins. Conversely, some genes exist in multiple copies. Therefore, the amount of protein sequence information in a cell cannot easily be estimated from the genome size.

2.1.1 Genes

A single gene coding for a particular protein corresponds to a sequence of nucleotides along one or more regions of a molecule of DNA. The DNA sequence is collinear with the protein sequence.

In species for which the genetic material is double-stranded DNA, genes may appear on either strand. Bacterial genes are continuous regions of DNA. Therefore, the functional unit of genetic sequence information from a bacterium is a string of $3N$ nucleotides encoding a string of N amino acids, or a string of N nucleotides encoding a structural RNA molecule of N residues. Such a string, equipped with annotations, would form a typical entry in one of the genetic sequence archives. In eukaryotes the nucleotide sequences that encode the amino acid sequences of individual proteins are organized in a more complex manner.

The relationship between size of gene and size of protein encoded is very different from that in bacteria. Frequently one gene appears split into separated segments in the genomic DNA. An *exon* is a stretch of DNA retained in the mature messenger RNA that the ribosome translates into protein. An *intron* is an intervening region between two exons. Cellular machinery splices together the proper segments, in RNA transcripts, based on signal sequences flanking the exons in the sequences themselves.

Many introns are very long—in some cases substantially longer than the exons. Regulatory mechanisms organize the expression of genes. Genes may be turned on or off (or more finely regulated) in response to concentrations of nutrients, or to stress, or to unfold complex programs of development during the lifetime of the organism.

Many control regions of DNA lie near the segments coding for proteins. They contain sequences that serve as binding sites for the molecules that transcribe the DNA sequence, or sequences that bind regulatory molecules that can *block* transcription. Simple examples occur in bacterial genomes, in which contiguous genes, coding for several proteins that catalyse successive steps in an integrated sequence of reactions, fall under the control of the same regulatory sequence. F. Jacob, J. Monod, and E. Wollman named these *operons*. One can readily understand the utility of a parallel mechanism for control of their expression.

In animals, methylation of DNA provides the signals for tissue-specific expression of developmentally regulated genes. Products of certain genes cause cells to commit suicide—a process called *apoptosis*. Defects in the apoptotic mechanism leading to uncontrolled growth are observed in some cancers, and stimulation of these mechanisms is a general approach to cancer therapy.

Table 2.1. Genome sizes.

Organism	*Number of Base pairs*	*Number of Genes*	*Comment*
ϕX-174	5386	10	virus infecting *E. coli*
Human mitochondrion	16569	37	subcellular organelle
Epstein-Barr virus (EBV)	172282	80	cause of mononucleosis
Mycoplasma pneumoniae	816394	680	cause of cyclic pneumonia epidemics
Rickettsia prowazeki	1111523	878	bacterium, cause of epidemic typhus
Treponema pallidum	1138011	1039	bacterium, cause of syphilis
Borrelia burgdorferi	1471725	1738	bacterium, cause of Lyme disease
Aquifex aeolicus	1551335	1749	bacterium from hot spring
Thermoplasma acidophilum	1564905	1509	archaeal prokaryote, lacks cell wall
Campylobacter jejuni	1641481	1708	frequent cause of food poisoning
Helicobacter pylori	1667867	1589	chief cause of stomach ulcers
Methanococcus jannaschii	1664970	1783	archaeal prokaryote, thermophile
Hemophilus influenzae	1830138	1738	bacterium, cause of middle ear infections
Thermotoga maritima	1860725	1879	marine bacterium
Archaeoglobus fulgidus	2178400	2437	another archaeon
Deinococcus radiodurans	3284156	3187	radiation-resistant bacterium
Synechocystis	3573470	4003	cyanobacterium, 'blue-green alga'
Vibrio cholerae	4033460	3890	cause of cholera
Mycobacterium tuberculosis	4411529	4275	cause of tuberculosis
Bacillus subtilis	4214814	4779	popular in molecular biology
Escherichia coli	4639221	4406	molecular biologists' all-time favourite
Pseudomonas aeruginosa	6264403	5570	largest prokaryote sequenced as yet
Saccharomyces cerevisiae	12.1×10^6	5885	yeast, first eukaryotic genome sequenced
Caenorhabditis elegans	95.5×10^6	19099	the worm
Arabidopsis thaliana	1.17×10^8	25498	flowering plant
Drosophila melanogaster	1.8×10^8	13601	the fruit fly

Fugu rubripes	3.9×10^8	30000	puffer fish (fugu fish)
Human	3.2×10^9	34000?	
Wheat	16×10^9	30000	
Salamander	10^{11}	?	
Psilotum nudum	10^{11}	?	whisk fern–a simple plant

The conclusion is that to reduce genetic data to individual coding sequences is to disguise the very complex nature of the interrelationships among them, and to ignore the historical and integrative aspects of the genome. Robbins has expressed the situation unimprovably:

... Consider the 3.2 gigabytes of a human genome as equivalent to 3.2 gigabytes of files on the mass-storage device of some computer system of unknown design. Obtaining the sequence is equivalent to obtaining an image of the contents of that mass-storage device. Understanding the sequence is equivalent to reverse engineering that unknown computer system (both the hardware and the 3.2 gigabytes of software) all the way back to a full set of design and maintenance specifications.

Reverse engineering the sequence is complicated by the fact that the resulting image of the mass-storage device will not be a file-by-file copy, but rather a streaming dump of the bytes in the order they were entered into the device. Furthermore, the files are known to be fragmented. In addition, some of the device contains erased files or other garbage. Once the garbage has been recognized and discarded and the fragmented files reassembled, the reverse engineering of the codes can be undertaken with only a partial, and sometimes incorrect, understanding of the CPU [Central Processing Unit] on which the codes run. In fact, deducing the structure and function of the CPU is part of the project, since some of the 3.2 gigabytes are the binary specifications for the computer-assisted-manufacturing process that fabricates the CPU. In addition, one must also consider that the huge database also contains code generated from the result of literally millions of maintenance revisions performed by the worst possible set of kludge-using, spaghetti-coding, opportunistic hackers who delight in clever tricks like writing self-modifying code and relying upon undocumented system quirks.

2.1.2 Proteins

In principle, a database of amino acid sequences of proteins is inherent in the database of nucleotide sequences of DNA, by virtue of the genetic code. Indeed, new protein sequence data are now being determined by translation of DNA sequences, rather than by direct sequencing of proteins. (Historically, the chemical problem of determining amino acid sequences of proteins directly was solved before the genetic code was established and before methods for determination of nucleotide sequences of DNA were developed. F. Sanger's sequencing of insulin in 1955 first proved that proteins had definite amino acid sequences, a proposition that until then was hypothetical.) Should any distinction be made between amino acid sequences determined directly from proteins and those determined by translation from DNA? First, we must assume that it is possible correctly to identify within the DNA data stream the regions that encode proteins.

The pattern-recognition programs that address this question are subject to three types of errors: a genuine protein sequence may be missed entirely, or an incomplete protein may be reported, or a gene may be incorrectly spliced. Several variations on the theme add to the complexity: Genes for different proteins may overlap, or genes may be assembled from exons in different ways in different tissues. Conversely, some genetic sequences that appear to code for proteins may, in fact, be defective or not expressed. *A protein inferred from a genome sequence is a hypothetical object until an experiment verifies its existence.* Second, in many cases the expression of a gene produces a molecule that must be modified within a cell, to make a *mature* protein that differs significantly from the one suggested by translation of the gene sequence.

In many cases the missing details of *post-translational modifications*—the molecular analogues of body piercing—are quite important, and unlike body piercing, have functional significance. Post-translational modifications include addition of ligands (for instance the covalently-bound haem group of cytochrome *c*), glycosylation, methylation, excision of peptides, and many others. Patterns of disulphide bridges—primary chemical bonds between cysteine residues—cannot be deduced from the amino acid sequence. In some cases, mRNA is edited before translation, creating changes in amino acid sequences that are not inferrable from the genes.

2.1.3 Proteomes

An organism's genome gives a complete but static set of specifications of the potential life of that individual. The state of development of the organism, and its activity at the molecular level at any moment, depend primarily on the amounts and distribution of its proteins. The *proteome project* is a large-scale programme dealing in an integral way with patterns of expression of proteins in biological systems, in ways that complement and extend genome projects. What kinds of data would we like to measure, and what mature experimental techniques exist to determine them? The basic goal is a spatio-temporal description of the deployment of proteins in the organism. The rates of synthesis of different proteins vary among different tissues and different cell types and states of activity.

Methods are available for efficient analysis of transcription patterns of multiple genes. However, because proteins 'turn over' at different rates, it is also necessary to measure proteins directly. The distribution of expressed protein levels is a kinetic balance between rates of protein synthesis and degradation. High-resolution two-dimensional polyacrylamide gel electrophoresis (2D PAGE) shows the pattern of protein content in a sample. Mass-spectroscopic techniques identify the proteins into which the sample has been separated, and their post-translational modifications. It is also possible to make arrays of proteins, to screen for pairs of proteins that interact. Application of these methods provides a picture of the protein-based activity of an organism, as the genome provides a complete set of potential proteins. R. Simpson has drawn the analogy: if the genome is a list of the instruments in an orchestra, the proteome is the orchestra in the process of playing a symphony.

2.2 EAVESDROPPING ON THE TRANSMISSION OF GENETIC INFORMATION

How hereditary information is stored, passed on, and implemented is perhaps *the* fundamental problem of biology. Three types of maps have been essential:

1. Linkage maps of genes
2. Banding patterns of chromosomes
3. DNA sequences

These represent three very different types of data. Genes, as discovered by Mendel, were entirely abstract entities. Chromosomes are physical objects, the banding patterns their visible landmarks. Only with DNA sequences are we dealing directly with stored hereditary information in its physical form. It was the very great achievement of the last century of biology to forge connections between these three types of data.

The first steps—and giant strides they were indeed—proved that, for any chromosome, the maps are one-dimensional arrays, and indeed that they are collinear. Any schoolchild now knows that genes are strung out along chromosomes, and that each gene corresponds to a DNA sequence. But the proofs of these statements earned a large number of Nobel prizes. Splitting a long molecule of DNA—for example, the DNA in an entire chromosome—into fragments of convenient size for cloning and sequencing requires additional maps to report the order of the fragments, so that the entire sequence can be reconstructed from the sequences of the fragments. A restriction endonuclease is an enzyme that cuts DNA at a specific sequence, usually about 6 bp long. Cutting DNA with several restriction enzymes with different specificities produces sets of overlapping fragments.

From the sizes of the fragments it is possible to construct a *restriction map*, stating the order and distance between the restriction enzyme cleavage sites. A mutation in one of these cleavage sites will change the sizes of the fragments produced by the corresponding enzyme, allowing the mutation to be located in the map. Restriction enzymes can produce fairly large pieces of DNA. Cutting the DNA into smaller pieces, which are cloned and ordered by sequence overlaps produces a finer dissection of the DNA called *contig map*.

In the past, the connections between chromosomes, genes and DNA sequences have been essential for identifying the molecular deficits underlying inherited diseases, such as Huntington disease or cystic fibrosis. Sequencing of the human genome has changed the situation radically. Nevertheless, Huntington disease and cystic fibrosis result from dysfunction of a single protein encoded by a single gene, and are not correlated with environmental factors or lifestyle. In contrast, many common diseases including diabetes and cancer are affected by many genes. Given a disease attributable to a defective protein:

1. If we know the protein involved, we can pursue rational approaches to therapy.
2. If we know the gene involved, we can devise tests to identify sufferers or carriers.
3. In many cases, knowledge of the chromosomal location of the gene is unnecessary for either therapy or detection; it is required only for identifying the gene, providing a bridge between the patterns of inheritance and the DNA sequence.

For instance, in the case of sickle-cell anaemia, we know the protein involved. The disease arises from a single point mutation in haemoglobin. We can proceed directly to drug design. We need the DNA sequence only for genetic testing and counselling. In contrast, if we know neither the protein nor the gene, we must somehow go from the phenotype back to the gene, a process called *positional cloning* or *reverse genetics*. Positional cloning used to involve a kind of Tinker to Evers to Chance cascade from the gene map to the chromosome map to the DNA sequence. Patterns of inheritance identify the type of genetic defect responsible for a condition. They show, for example, that Huntington disease and cystic fibrosis are caused by single genes.

To find the gene associated with cystic fibrosis it was necessary to begin with the gene map, using linkage patterns of heredity in affected families to localize the affected gene to a particular region of a particular chromosome. Knowing the general region of the chromosome, it was then possible to search the DNA of that region to identify candidate genes, and finally to pinpoint the particular gene responsible and sequence it. In contrast, many diseases do not show simple inheritance, or, even if only a single gene is involved, heredity creates only a predisposition the clinical consequences of which depend on environmental factors. The full human genome sequence, and measurements of expression patterns, will be essential to identify the genetic components of these more complex cases.

2.2.1 Mappings between the Maps

A gene linkage map can be calibrated to chromosome banding patterns through observation of individuals with deletions or translocations of parts of chromosomes. The genes responsible for phenotypic changes associated with a deletion must lie within the deletion. Translocations are correlated with altered patterns of linkage and recombination. There have been several approaches to coordinating chromosome banding patterns with individual DNA sequences of genes:

1. In Fluorescent *In Situ* Hybridization (FISH) a probe sequence is labelled with fluorescent dye. The probe is hybridized with the chromosomes, and the chromosomal location where the probe is bound shows up directly in a photograph. Typical resolution is $\sim 10^5$ base pairs, but specialized new techniques can achieve high resolution, down to 1 kbp. Simultaneous FISH with two probes can detect linkage and even estimate genetic distances. This is important in species for which the generation

time is long enough to make standard genetic approaches inconvenient. FISH can also detect chromosomal abnormalities.

2. *Somatic cell hybrids* are rodent cells containing few, one, or even partial human chromosomes. (Chromosome fragments are produced by irradiating the human cells prior to fusion. Such lines are called *radiation hybrids*.) Hybridization of a probe sequence with a panel of somatic cell hybrids, detected by fluorescence, can identify which chromosome contains the probe. This approach has been superseded by use of clones of yeast or bacteria or phage containing fragments of human DNA in artificial chromosomes (YACs and BACs and PACs).

Of course, for human DNA these methods are obsolete, now that the sequence is known (but mind the gaps). Given a DNA sequence one would just look it up.

2.2.2 High-resolution Maps

Formerly, genes were the only visible portions of genomes. Now, markers are no longer limited to genes with phenotypically observable effects, which are anyway too sparse for an adequately high-resolution map of the human genome. Now that we can interrogate DNA sequences directly, any features of DNA that vary among individuals can serve as markers, including:

1. *Variable number tandem repeats* (VNTRs), also called minisatellites. VNTRs contain regions 10-100 base pairs long, repeated a variable number of times-same sequence, different number of repeats. In any individual, VNTRs based on the same repeat motif may appear only once in the genome; or several times, with different lengths on different chromosomes. The distribution of the sizes of the repeats is the marker. Inheritance of VNTRs can be followed in a family and mapped to a disease phenotype like any other trait. VNTRs were the first genetic sequence data used for personal identification—genetic fingerprints—in paternity and in criminal cases.

   Formerly, VNTRs were observed by producing *restriction fragment length polymorphisms* (RFLPs) from them. VNTRs are generally flanked by recognition sites for the same restriction enzyme, which will neatly excise them. The results can be spread out on a gel, and detected by Southern blotting. Note the distinction: VNTRs are characteristics of genome sequences; RFLPs are artificial mixtures of short stretches of DNA created in the laboratory in order to identify VNTRs.

It is much easier and more efficient to measure the sizes of VNTRs by amplifying them with PCR, and this method has replaced the use of restriction enzymes.

2. *Short tandem repeat polymorphisms* (STRPs), also called microsatellites. STRPs are regions of only 2-5 bp but repeated many times; typically 10-30 consecutive copies. They have several advantages as markers over VNTRs, one of which is a more even distribution over the human genome.

There is no reason why these markers need lie within expressed genes, and usually they do not. (The CAG repeats in the gene for huntingtin and certain other disease genes are exceptions.) Panels of micro satellite markers greatly simplify the identification of genes.

It is interesting to compare a recent project to identify a disease gene now that the human genome sequence is available, with such classic studies as the identification of the gene for cystic fibrosis. Additional mapping techniques deal more directly with the DNA sequences, and can short-circuit the process of gene identification:

1. A *contig*, or *contiguous clone map*, is a series of overlapping DNA clones of known order along a chromosome from an organism of interest—for instance, human—stored in yeast or bacterial cells as YACs (Yeast Artificial Chromosomes) or BACs (Bacterial Artificial Chromosomes). A contig map can produce a very fine mapping of a genome. In a YAC, human DNA is stably integrated into a small extra chromosome in a yeast cell. A YAC can contain up to 10^6 base pairs. In principle, the entire human genome could be represented in 10 000 YAC clones. In a BAC, human DNA is inserted into a plasmid in an *E. coli* cell. (A plasmid is a small piece of double-stranded DNA found in addition to the main genome, usually but not always circular.) A BAC can carry about 250 000 bp. Despite their smaller capacities, BACs are preferred to YACs because of their greater stability and ease of handling.

2. *A sequence tagged site* (STS) is a short, sequenced region of DNA, typically 200-600 bp long, that appears in a unique location in the genome. It need not be polymorphic. An STS can be mapped into the genome by using PCR to test for the presence of the sequence in the cells containing a contig map.

One type of STS arises from an *expressed sequence tag* (EST), a piece of cDNA (complementary DNA; i.e., a DNA sequence derived from the messenger RNA of an expressed gene.) The sequence contains only the exons of the gene, spliced together to form the sequence that encodes the protein. cDNA sequences can be mapped to chromosomes using FISH, or located within contig maps. How do contig maps and sequence tagged sites facilitate identifying genes? If you are working with an organism for which the full genome sequence is not known, but for which full contig maps are available for all chromosomes, you would identify STS markers tightly linked to your gene, then locate these markers in the contig maps.

2.3 PICKING OUT GENES IN GENOMES

Computer programs for genome analysis identify *open reading frames* or *ORFs*. An ORF is a region of DNA sequence that begins with an initiation codon (ATG) and ends with a stop codon. An ORF is a potential protein-coding region. Approaches to identifying protein-coding regions choose from or combine two possible approaches:

1. *Detection of regions similar to known coding regions from other organisms*. These regions may encode amino acid sequences similar to known proteins, or may be similar to Espressed Sequence Tags (ESTs). Because ESTs are derived from messenger RNA, they correspond to genes known to be expressed. It is necessary to sequence only a few hundred initial bases of cDNA to give enough information to identify a gene: characterization of genes by ESTs is like indexing poems or songs by their first lines.
2. *Ab initio methods, that seek to identify genes from the properties of the DNA sequences themselves*. Computer-assisted annotation of genomes is more complete and accurate for bacteria than for eukaryotes. Bacterial genes are relatively easy to identify because they are contiguous—they lack the introns characteristic of eukaryotic genomes—and the intergene spaces are small. In higher organisms, identifying genes is harder. Identification of exons is one problem, assembling them is another. Alternative splicing patterns present a particular difficulty.

A framework for ab initio gene identification in eukaryotic genomes includes the following features:

1. The initial (5') exon starts with a transcription start point, preceded by a core promotor site such as the TATA box typically ~30bp upstream. It is free of in-frame stop codons, and ends immediately before a GT splice signal. (Occasionally a non-coding exon precedes the exon that contains the initiator codon.)

2. Internal exons, like initial exons, are free of in-frame stop codons. They begin immediately after an AG splice signal and and end immediately before a GT splice signal. The flanking introns contain consensus branch point sequences and polypyrimidine tracts that interact with the splicing apparatus.
3. The final (3') exon starts immediately after an AG splice signal and ends with a stop codon, followed by a polyadenylation signal sequence. (Occasionally a noncoding exon follows the exon that contains the stop codon.)

All coding regions have non-random sequence characteristics, based partly on codon usage preferences. Empirically, it is found that statistics of hexanucleotides perform best in distinguishing coding from non-coding regions. Starting from a set of known genes from an organism as a training set, pattern recognition programs can be tuned to particular genomes. Accurate gene detection is a crucial component of genome sequence analysis.

2.4 GENOMES OF PROKARYOTES

Most prokaryotic cells contain their genetic material in the form of a large single circular piece of double-stranded DNA, usually less than 5 Mb long. In addition they may contain plasmids. The protein-coding regions of bacterial genomes do not contain introns.

In many prokaryotic genomes the protein-coding regions are partially organized into *operons*—tandem genes transcribed into a single messenger RNA molecule, under common transcriptional control. In bacteria, the genes of many operons code for proteins with related functions. For instance, successive genes in the trp operon of *E. coli* code for proteins that catalyse successive steps in the biosynthesis of tryptophan. In archaea, a metabolic relationship between genes in operons is less frequently observed. The typical prokaryotic genome contains only a relatively small amount of non-coding DNA (in comparison with eukaryotes), distributed throughout the sequence. In *E. coli* only ~11 % of the DNA is non-coding.

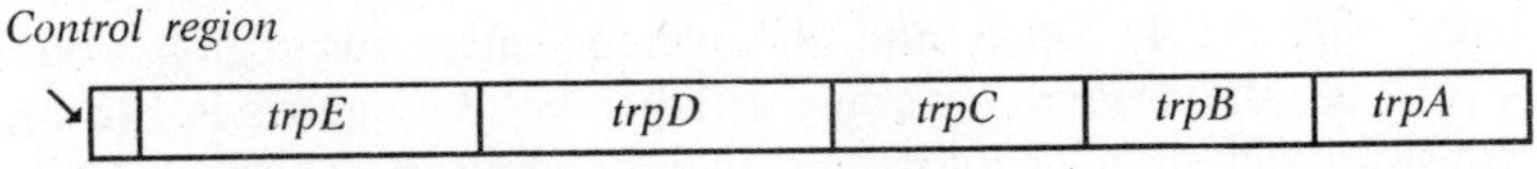

Figure 2.1. The trp operon in *E. coli*.

2.4.1 Genome of the Bacterium *Escherichia coli*

E. coli, strain K-12, has long been the workhorse of molecular biology. The genome of strain MG1655, contains 4 639 221 bp in a single circular DNA molecule, with no plastids. Approximately 89% of the sequence codes for proteins or structural RNAs. An inventory reveals:

- 4285 protein-coding genes
- 122 structural RNA genes
- non-coding repeat sequences
- regulatory elements
- transcription/translation guides
- transposases
- prophage remnants
- insertion sequence elements
- patches of unusual composition, likely to be foreign elements introduced by horizontal transfer.

Analysis of the genome sequence required identification and annotation of protein-coding genes and other functional regions. Many *E. coli* proteins were known before the sequencing was complete, from the many years of intensive investigation: 1853 proteins had been described before publication of the genome sequence. Other genes could be assigned functions from identification of homologues by searching in sequence databanks. The narrower the range of specificity of the function of the homologues, the more precise could be the assignment. Currently, over 60% of proteins can be assigned at least a general function. Other regions of the genome are recognized as regulatory sites, or as mobile genetic elements, also on the basis of similarity to homologous sequences known in other organisms.

The distribution of protein-coding genes over the genome of *E. coli* does not seem to follow any simple rules, either along the DNA or on different strands. Indeed, comparison of strains suggests that genes are mobile. The *E. coli* genome is relatively gene dense. Genes coding for proteins or structural RNAs occupy ~89% of the sequence. The average size of an ORF is 317 amino acids. If the genes were evenly distributed, the average intergenic region would be 130 bp; the observed average distance between genes is 118 bp. However, the sizes of intergenic regions vary considerably. Some intergenic regions are large. These contain sites of regulatory function, and repeated sequences. The longest intergenic region, 1730 bp, contains non-coding repeat sequences. Approximately three-fourth of the transcribed units contain only one gene; the rest contain several consecutive genes, or operons. It is estimated that the *E. coli* genome contains 630—700 operons. Operons vary in size, although few contain more than five genes. Genes within operons tend to have related functions.

Table 2.2. Distribution of *E. coli* proteins among 22 functional groups.

Functional class	*Number*	*%*
Regulatory function	45	1.05
Putative regulatory proteins	133	3.10
Cell structure	182	4.24
Putative membrane proteins	13	0.30
Putative structural proteins	42	0.98
Phage, transposons, plasmids	87	2.03
Transport and binding proteins	281	6.55
Putative transport proteins	146	3.40
Energy metabolism	243	5.67
DNA replication, recombination, modification, and repair	115	2.68
Transcription, RNA synthesis, metabolism, and modification	55	1.28
Translation, post-translational protein modification	182	4.24
Cell processes (including adaptation, protection)	188	4.38
Biosynthesis of cofactors, prosthetic groups, and carriers	103	2.40
Putative chaperones	9	0.21
Nucleotide biosynthesis and metabolism	58	1.35
Amino acid biosynthesis and metabolism	131	3.06
Fatty acid and phospholipid metabolism	48	1.12
Carbon compound catabolism	130	3.03
Central intermediary metabolism	188	4.38
Putative enzymes	251	5.85
Other known genes (gene product or phenotype known)	26	0.61
Hypothetical, unclassified, unknown	1632	38.06

In some cases, the same DNA sequence encodes parts of more than one polypeptide chain. One gene codes for both the τ and γ subunits of DNA polymerase III. Translation of the entire gene forms the τ subunit. The γ subunit corresponds approximately to the N-terminal two-thirds of the τ subunit. A frame shift on the ribosome at this point leads to chain termination 50% of the time, causing a 1: 1 ratio of expressed τ and γ subunits. There do not appear to be any overlapping genes in which different reading frames both code for expressed proteins. In other cases, the same polypeptide chain appears in more than one enzyme. A protein that functions on its own as lipoate dehydrogenase is also an essential subunit of pyruvate dehydrogenase, 2-oxoglutarate dehydrogenase and the glycine cleavage complex. Having the complete genome, we can examine the protein repertoire of *E. coli.* The largest class of proteins are the enzymes —approximately 30% of the total genes. Many enzymatic functions are shared by more than one protein.

Some of these sets of functionally similar enzymes are very closely related, and appear to have arisen by duplication, either in *E. coli* itself or in an ancestor. Other sets of functionally similar enzymes have very dissimilar sequences, and differ in specificity, regulation, or intracellular location. Several features of *E. coli's* generous endowment of enzymes give it an extensive and versatile metabolic competence, which allow it to grow and compete under varying conditions:

1. It can synthesize all components of proteins and nucleic acids (amino acids and nucleotides), and cofactors.
2. It has metabolic flexibility: Both aerobic and anaerobic growth are possible, utilizing different pathways of energy capture. It can grow on many different carbon and nitrogen sources. Not all metabolic pathways are active at any time; the alternatives allow response to changes in conditions.
3. A wide range of transporters allows uptake of required substrates.
4. Even for specific metabolic reactions there are many cases of multiple enzymes. These provide redundancy, and contribute to an ability to tune metabolism to varying conditions. Complex regulatory mechanisms integrate the protein expression pattern.
5. However, *E. coli* does not possess a complete range of enzymatic capacity. It cannot fix CO_2 or N_2.

We have described here some of the static features of the *E. coli* genome and its protein repertoire. Current research is moving into dynamic aspects, to investigations of protein expression patterns.

2.4.2 Genome of the Archaeon *Methanococcus jannaschii*

S. Luria once suggested that to determine common features of all life one should not try to survey everything, but rather identify the organism most different from us and see what we have in common with it. The assumption was that the way to do this would be to find an organism adapted to the most different environment. Deep-sea exploration has revealed environments as far from the familiar as those portrayed in science fiction.

Hydrothermal vents are underwater volcanoes emitting hot lava and gases through cracks in the ocean floor. They create niches for communities of living things disconnected from the surface, depending on the minerals exuded from the vent for their sources of inorganic nutrients. They support living communities of microorganisms that are the only known forms of life not dependent on sunlight, directly or indirectly, for their energy source. The microorganism *Methanococcus jannaschii* was collected from a hydrothermal vent 2600 m deep off the coast of Baja California, Mexico, in 1983. It is a thermophilic organism, surviving at temperatures from 48—94°C, with an optimum at 85°C. It is a strict anaerobe, capable of self-reproduction from inorganic components. Its overall metabolic equation is to synthesize methane from H_2 and CO_2.

M. jannaschii belongs to the archaea, one of the three major divisions of life along with the bacteria and eukarya. The archae a comprise groups of prokaryotes, including organisms adapted to extreme environmental conditions such as high temperature and pressure, or high salt concentration. The genome of *M. jannaschii* was sequenced in 1996 by The Institute for Genomic Research (TIGR). It was the first archaeal genome sequenced. It contains a large chromosome containing a circular double-stranded DNA molecule 1 664 976 bp long, and two extrachromosomal elements of 58 407 and 16 550 bp, respectively.

There are 1743 predicted coding regions, of which 1682 are on the chromosome, and 44 and 12 on the large and small extrachromosomal elements respectively. Some RNA genes contain introns. As in other prokaryotic genomes there is little non-coding DNA. *M. jannaschii* would appear to satisfy Luria's goal of finding our most distant extant relative. Comparison of its genome sequence with others shows that it is distantly related to other forms of life. Only 38% of the ORFs could be assigned a function, on the basis of homology to proteins known from other organisms. (Now over 50% can be assigned a function.) However, to everyone's great surprise, archaea are in some ways more closely related to eukaryotes than to bacteria! They are a complex mixture. In archaea, proteins involved in transcription, translation, and regulation are more similar to those of eukaryotes. Archaeal proteins involved in metabolism are more similar to those of bacteria.

2.4.3 Genome of one of the simplest organisms: *Mycoplasma genitalium*

Mycoplasma genitalium is an infectious bacterium, the cause of non-gonococcal uretheitis. Its genome was sequenced in 1995 by a collaboration of groups at TIGR, The Johns Hopkins University and The University of North Carolina. The genome is a single DNA molecule containing 580 070 bp. This is the smallest cellular genome yet sequenced. So far, *M. genitalium* is the closest we have to a *minimal organism,* the smallest capable of independent life. (Viruses, in contrast, require the cellular machinery of their hosts.) The genome is dense in coding regions. Four hundred and sixty-eight genes have been identified as expressed proteins. Some regions of the sequence are gene rich, others gene poor, but overall 85% of the sequence is coding.

The average length of a coding region is 1040 bp. As in other bacterial genomes, the coding regions do not contain introns. Further compression of the genome is achieved by overlapping genes. It appears that many of these have arisen through loss of stop codons. The gene repertoire of *M. genitalium* includes some that encode proteins essential for independent reproduction, such as those involved in DNA replication, transcription, and translation, plus ribosomal and transfer RNAs. Other genes are specific for the infectious activity. These include adhesins that mediate binding to infected cells, other molecules for defence against the host's immune system, and many transport proteins. As adaptation to the parasitic lifestyle of the organism, there has been widespread loss of metabolic enzymes, including those responsible for amino acid biosynthesis—indeed, one amino acid is absent from all *M. genitalium* proteins.

2.5 GENOMES OF EUKARYOTES

It is rare in science to encounter a completely new world containing phenomena entirely unsuspected. The complexity of the eukaryotic genome is such a world. In eukaryotic cells, the majority of DNA is in the nucleus, separated into bundles of nucleoprotein, the chromosomes. Each chromosome contains a single double-stranded DNA molecule. Smaller amounts of DNA appear in organelles—mitochondria and chloroplasts. The organelles originated as intracellular parasites.

Organelle genomes usually have the form of circular double-stranded DNA, but are sometimes linear and sometimes appear as multiple circles. The genetic code by which organelle genes are translated differs from that of nuclear genes. Nuclear genomes of different species vary widely in size. The correlation between genome size and complexity of the organism is very rough. It certainly does not support any preconception that humans stand on a pinnacle. In many cases differences in genome size reflect different amounts of simple repetitive sequences, often referred to as 'junk DNA.' In addition to variation in DNA content, eukaryotic species vary in the number of chromosomes and distribution of genes among them.

Some differences in the distribution of genes among chromosomes involve translocations, or chromosome fragmentations or joinings. For instance, humans have 23 pairs of chromosomes; chimpanzees have 24. Human chromosome 2 is equivalent to a fusion of chimpanzee chromosomes 12 and 13. The difficulty of chromosome pairing during mitosis in a zygote after such an event can contribute to the reproductive isolation required for species separation. Other differences in chromosome complement reflect duplication or hybridization events. The wheat first used in agriculture, in the Middle East at least 10 000—15 000 years ago, is a diploid called *einkorn* (*Triticum monococcum*), containing 14 pairs of chromosomes. *Emmer* wheat (*T. dicoccum*), also cultivated since palaeolithic times, and *durum* wheat (*T. turgidum*), are merged hybrids of relatives of einkorn with other wild grasses, to form tetraploid species.

Additional hybridizations, to different wild wheats, gave hexaploid forms, including *spelt* (*T. spelta*), and modern common wheat *T. aestivum*. *Triticale*, a robust crop developed in modern agriculture and currently used primarily for animal feed, is an artificial genus arising from crossing durum wheat (*Triticum turgidim*) and rye (*Secale cereale*). Most triticale varieties are hexaploids.

Variety of wheat	*Classification*	*Chromosome complement*
Einkorn	*Triticum monococcum*	AA
emmer wheat	*Triticum turgidum*	AABB
durum wheat	*Triticum turgidum*	AABB
spelt	*Triticum spelta*	AABBDD
common wheat	*Triticum aestivum*	AABBDD
triticale	*Triticosecale*	AABBRR

All these species are still cultivated—some to only minor extents—and have their individual uses in cooking. Spelt, or *farro* in Italian, is the basis of a well-known soup; pasta is made from durum wheat; and bread from *T. aestivum*. Even within single chromosomes, gene families are common in eukaryotes. Some family members are *paralogues*—related genes that have duplicated within the same genome, and have in many cases diverged to provide separate functions in descendant species. Changes in expression pattern may precede development of novel function. (*Orthologues*, in contrast, are homologues in different species that often perform the same function. For instance, human a and b globin are paralogues, and human and horse myoglobin are orthologues.) Other related sequences may be pseudogenes, which may have arisen by duplication, or by retrotransposition from messenger RNA, followed by the accumulation of mutations to the point of loss of function. The human globin gene cluster is a good example.

2.5.1 Genome of *Saccharomyces cerevisiae* (Baker's yeast)

Yeast is one of the simplest known eukaryotic organisms. Its cells, like our own, contain a nucleus and other specialized intracellular compartments. The sequencing of its genome, by an unusually effective international consortium involving ~100 laboratories, was completed in 1992. The yeast genome contains 12 057 500 bp of nuclear DNA, distributed over 16 chromo-somes. The chromosomes range in size over an order of magnitude, from the 1352 kbp chromosome IV to the 230 kbp chromosome I. The yeast genome contains 5885 predicted protein-coding genes, ~140 genes for ribosomal RNAs, 40 genes for small nuclear RNAs, and 275 transfer RNA genes.

In two respects, the yeast genome is denser in coding regions than the known genomes of the more complex eukaryotes *Caenorhabditis elegans*, *Drosophila melanogaster*, and human: (1) Introns are relatively rare, and relatively small. Only 231 genes in yeast contain introns. (2) There are fewer repeat sequences compared with more complex eukaryotes. A duplication of the entire yeast genome appears to have occurred ~150 mil. lion years ago. This was followed by translocations of pieces of the duplicated DNA and loss of one of the copies of most (~92%) of the genes. Of the 5885 potential protein-coding genes, 3408 correspond to known proteins.

About 1000 more contain some similarity to known proteins in other species. Another ~800 are similar to ORFs in other genomes that correspond to unknown proteins. Many of these homologues appear in prokaryotes. Only approximately one-third of yeast proteins have identifiable homologues in the human genome. In taking censuses of genes, it has been useful to classify their functions into broad categories. The following classification of yeast protein functions is taken from http://www.mips.biochem. mpg.de/proj/yeast/catalogues/funcat/:

- Metabolism
- Energy
- Cell growth, cell division and DNA synthesis
- Transcription
- Protein synthesis
- Protein destination
- Transport facilitation
- Cellular transport and transport mechanisms
- Cellular biogenesis
- Cellular communication/signal transduction
- Cell rescue, defence, cell death and ageing
- Ionic homeostasis
- Cellular organization
- Transposable elements, viral and plasmid proteins
- Unclassified

Yeast is a testbed for development of methods to assign functions to gene products. The search for homologues has been exhaustive and it continues. Collections of mutants exist containing a knockout of every gene. (A unique sequence 'bar code' introduced into each mutant facilitates identification of the ones that grow under selected conditions.) Cellular localization and expression patterns are being investigated. Several types of measurements, including those based on activation of transcription by pairs of proteins that can form dimers, are producing catalogues of interprotein interactions.

2.5.2 Genome of *Caenorhabditis elegans*

The nematode worm *Caenorhabditis elegans* entered biological research in the 1960s, at the express invitation of Sydney Brenner. He recognized its potential as a sufficiently complex organism to be interesting but simple enough to permit complete analysis, at the cellular level, of its development and neural circuitry. The *C. elegans* genome, completed in 1998, provided the first full DNA sequence of a multicellular organism.

The *C. elegans* genome contains ~97 Mbp of DNA distributed on paired chromosomes I, II, III, IV, V and X. There is no Y chromosome. Different genders in *C. elegans* appear in the XX genotype, a self-fertilizing hermaphrodite; and the XO genotype, a male. The *C. elegans* genome is about eight times larger than that of yeast, and its 19,099 predicted genes are approximately three times the number in yeast.

Table 2.3. Distribution of *C. elegans* genes

Chromosome	*Size (Mb)*	*Number of progein genes*	*Density of protein genes (kb/genes)*	*Number of tRNA genes*
I	7.9	2803	5.06	13
II	8.5	3259	3.65	6
II	7.6	2508	5.40	9
IV	9.2	3094	5.17	7
V	9.8	4082	4.15	5
X	10.1	2631	6.54	3

The gene density is relatively low for a eukaryote, with ~1 gene/5 kb of DNA. Exons cover ~27% of the genome; the genes contain an average of 5 introns each. Approximately 25% of the genes are in clusters of related genes. Many *C. elegans* proteins are common to other life forms. Others are apparently specific to nematodes. Forty-two percent of proteins have homologues outside the phylum, 34% are homologous to proteins of other nematodes and 24% have no known homologues outside *C. elegans* itself. Many of the proteins have been classified according to structure and function.

Several kinds of RNA genes have been identified. The *C. elegans* genome contains 659 genes for tRNA, almost half of them (44%) on the X chromosome. Spliceosomal RNAs appear in dispersed copies, often identical. (Spliceosomes are the organelles that convert pre-mRNA transcripts to mature mRNA by excising introns.) Ribosomal RNAs appear in a long tandem array at the end of chromosome I. 5S RNAs appear in a tandem array on chromosome V. Some RNA genes appear in introns of protein-coding genes.

The *C. elegans* genome contains many repeat sequences. Approximately 2.6% of the genome consists of tandem repeats. Approximately 3.6% of the genome contains inverted repeats; these appear preferentially within introns, rather than between genes.

Repeats of the hexamer sequence TTAGGC appear in many places. There are also simple duplications, involving hundreds to tens of thousands of kilobases.

Table 2.4. *C. elegans*: 20 commonest protein domains

Type of domain	*Number*
Seven-transmembrane spanning chemoreceptor	650
Eukaryotic protein kinase domain	410
Two domain, C4 type zinc finger	240
Collagen	170
Seven-transmembrane spanning receptor (rhodopsin-family)	140
C2H2 type zinc finger	130
C-type lectin	120
RNA recognition motif	100
C3HC4 type (RING finger) zinc fingers	90
Protein tyrosine phosphatase	90
Ankyrin repeat	90
WD domain G-beta repeat	90
Homeobox domain	80
Neurotransmitter gated ion channel	80
Cytochrome P450	80
Conserved C-terminal helicase	80
Short chain and alcohol dehydrogenases	80
UDP-glucoronosyl and UDP-glucosyl transferases	70
EGF-like domain	70
Immunoglobulin superfamily	70

2.5.3 Genome of *Drosophila melanogaster*

Drosophila melanogaster, the fruit fly, has been the subject of detailed studies of genetics and development for almost a century. Its genome sequence, the product of a collaboration between Celera Genomics and the Berkeley Drosophila Genome Project, was announced in 1999. The chromosomes of *D. melanogaster* are nucleoprotein complexes. Approximately one-third of the genome is contained in heterochromatin, highly coiled and compact (and therefore densely staining) regions flanking the centromeres.

The other two-thirds is euchromatin, a relatively uncoiled, less-compact form. Most of the active genes are in the euchromatin. The heterochromatin in *D. melanogaster* contains many tandem repeats of the sequence AATAACATAG, and relatively few genes.

The total chromosomal DNA of *D. melanogaster* contains about ~180 Mbp. The released sequence consists of the euchromatic portion, about 120 Mbp.

The genome is distributed over five chromosomes: three large autosomes, a Y chromosome, and a fifth tiny chromosome containing only ~1 Mb of euchromatin. The fly's 13 601 genes are approximately double the number in yeast, but are fewer than in *C. elegans*, perhaps a surprise. The average density of genes in the euchromatin sequence is 1 gene/9 kb; much lower than the typical 1 gene/kb densities of prokaryotic genomes. Despite the fact that insects are not very closely related to mammals, the fly genome is useful in the study of human disease. It contains homologues of 289 human genes implicated in various diseases, including cancer, and cardiovascular, neurological, endocrinological, renal, metabolic, and haematological diseases. Some of these homologues have different functions in humans and flies.

Other human disease-associated genes can be introduced into, and studied in, the fly. For instance, the gene for human spinocerebellar ataxia type 3, when expressed in the fly, produces similar neuronal cell degeneration. There are now fly models for Parkinson disease and malaria. The non-coding regions of the *D. melanogaster* genome must contain regions controlling spatiotemporal patterns of development. The developmental biology of the fly has been studied very intensively. It is, therefore, an organism in which the study of the genomics of development will prove extremely informative.

2.5.4 Genome of *Arabidopsis thaliana*

As a flowering plant *Arabidopsis thaliana* is a very distant relative of the other higher eukaryotic organisms for which genome sequences are available. It invites comparative analysis to identify common and specialized features. The *Arabidopsis thaliana* genome contains ~125 Mbp of DNA, of which 115.4 Mbp was reported in 2000 by The *Arabidopsis* Genome Initiative.

There are five pairs of chromosomes, containing 25 498 predicted genes. The genome is relatively compact, with 1 gene/4.6 kb on the average. This figure is intermediate between prokaryotes and *Drosophila*, and roughly similar to *C. elegans*. The genes of *Arabidopsis* are relatively small. Exons are typically 250 bp long, and introns relatively small, with mean length 170 bp. Typical of plant genes is an enrichment of coding regions in GC content. Most

Arabidopsis proteins have homologues in animals, but some systems are unique, among higher organisms, to plants, including cell wall production and photosynthesis. Many proteins shared with animals have diverged widely since the last common ancestor.

Typical of another difference between plants and animals, 25% of the nuclear genes in plants have signal sequences governing their transport into organelles —mitochondria and chloroplasts — compared to 5% of mitochondrial-targeted nuclear genes in animals. Traces of large- and small-scale duplication events appear in the *Arabidopsis* genome. Fifty-eight per cent of the genome contains 24 duplicated segments, ≥100 kb in length. The genes in the duplicated segments are usually not identifiable homologues. However, 4140 genes (17% of all genes) appear in 1528 tandem arrays of gene families, containing up to 23 members.

The pattern of duplications suggests that an ancestor of *Arabidopsis* underwent a whole-genome duplication, to form a tetraploid species. That this happened a long time—estimated at 112 million years—ago is shown by the very high amount of loss of one copy of duplicated genes, and wide divergence of those that remain. Higher plants must integrate the effects of three genomes—nuclear, chloroplast and mitochondrial. The organelle genomes are much smaller:

2.5. Gene distribution in *A. thaliana* between nucleus and organelles

	Nuclear	*Chloroplast*	*Mitochondrion*
Size (kb)	125 100	154	367
Protein genes	25 498	79	58
Density (kb/protein gene)	4.5	1.2	6.25

Many genes for proteins synthesized by nuclear genes and transported to organelles appear to have originated in the organelles and been transferred to the nucleus.

2.6 GENOME OF *HOMO SAPIENS* (THE HUMAN GENOME)

In February 2001, the International Human Genome Sequencing Consortium and Celera Genomics published, separately, drafts of the human genome. The sequence amounted to $\sim 3.2 \times 10^9$ bp, thirty times larger than the genomes of *C. elegans* or *D. melanogaster*. One reason for this disparity in size is that coding sequences form less than 5% of the human genome; repeat sequences

2.6. The top categories in a functional classification.

Function	*Number*	*%*
Nucleic acid binding	2207	14.0
DNA binding	1656	10.5
DNA repair protein	45	0.2
DNA replication factor	7	0.0
Transcription factor	986	6.2
RNA binding	380	2.4
Structural protein of ribosome	137	0.8
Translation factor	44	0.2
Transcription factor binding	6	0.0
Cell cycle regulator	75	0.4
Chaperone	154	0.9
Motor	85	0.5
Actin binding	129	0.8
Defense/immunity protein	603	3.8
Enzyme	3242	20.6
Peptidase	457	2.9
Endopeptidase	403	2.5
Protein kinase	839	5.3
Protein phosphatase	295	1.8
Enzyme activator	3	0.0
Enzyme inhibitor	132	0.8
Apoptosis inhibitor	28	0.1
Signal transduction	1790	11.4
Receptor	1318	8.4
Transmembrane receptor	1202	7.6
G-protein linked receptor	489	3.1
Olfactory receptor	71	0.4
Storage protein	7	0.0
Cell adhesion	189	1.2
Structural protein	714	4.5
Cytoskeletal structural protein	145	0.9
Transporter	682	4.3
Ion channel	269	1.7
Neurotransmitter transporter	19	0.1
Ligand binding or carrier	1536	9.7
Electron transfer	33	0.2
Cytochrome P450	50	0.3
Tumour suppressor	5	0.0
Unclassified	4813	30.6
Total	15683	100.0

over 50%. Perhaps the most surprising feature was the small number of genes identified. The finding of only about 30 000-40 000 genes suggests that alternative splicing patterns make a very significant contribution to our protein repertoire. It is estimated that ~35% of genes have alternative splicing patterns. The human genome is distributed over 22 chromosome pairs plus the X and Y chromosomes.

The DNA contents of the autosomes range from 279 Mbp down to 48 Mbp. The X chromosome contains 163 Mbp and the Y chromosome only 51 Mbp. The exons of human protein-coding genes are relatively small compared to those in other known eukaryotic genomes. The introns are relatively long. As a result many protein-coding genes span long stretches of DNA. For instance, the dystrophin gene, coding for a 3 685 amino acid protein, is >2.4 Mbp long.

2.6.1 Protein Coding Genes

Analysis of the human protein repertoire implied by the genome sequence has proved difficult because of the problems in reliably detecting genes, and alternative splicing patterns. The International Human Genome Sequencing Consortium has estimated that the total count will be ~32000 genes in all. A classification based on structure revealed the most common types:

2.6.2 Repeat Sequences

Repeat sequences comprise over 50% of the genome:

- transposable elements, or interspersed repeats — almost half the entire genome! These include the LINEs and SINEs.
- retroposed pseudogenes
- simple 'stutters' —repeats of short oligomers. These include the minisatellites and microsatellites. Trinucleotide repeats such as CAG, corresponding to glutamine repeats in the corresponding protein, are implicated in numerous diseases.
- segmental duplications, of blocks of ~10-300 kb. Interchromosomal duplications appear on non-homologous chromosomes, sometimes at multiple sites. Some intrachromosomal duplications include closely spaced duplicated regions many kilobases long of very similar sequence implicated in genetic diseases; for example Charcot-Marie-Tooth syndrome type 1A, a progressive peripheral neuropathy resulting from duplication of a region containing the gene for peripheral myelin protein 22.
- blocks of tandem repeats, including gene families

2.7. Most common types of proteins

Protein	*Number*
Immunoglobulin and major histocompatibility complex domain	591
Zinc finger, C2H2 type	499
Eukaryotic protein kinase	459
Rhodopsin-like GPCR superfamily	346
Serine/Threonine protein kinase family active site	285
EGF-like domain	259
RNA-binding region RNP-1 (RNA recognition motif)	214
G-protein beta WD-40 repeats	196
Src homology 3 (SH3) domain	194
Pleckstrin homology (PH) domain	188
EF-hand family	185
Homeobox domain	179
Tyrosine kinase catalytic domain	173
Immunoglobulin V-type	163
RING finger	159
Proline rich extensin	156
Fibronectin type III domain	151
Ankyrin-repeat	135
KRAB box	133
Immunoglobulin subtype	128
Cadherin domain	118
PDZ domain (also known as DHR or GLGF)	117
Leucine-rich repeat	113
Serine proteases, trypsin family	108
Ras GTPase superfamily	103
Src homology 2 (SH2) domain	100
BTB/POZ domain	99
TPR repeat	92
AAA ATPase superfamily	92
Aspartic acid and asparagine hydroxylation site	91

2.6.3 RNA

RNA genes in the human genome include:

1. 497 transfer RNA genes. One large cluster contains 140 tRNA genes within a 4 Mb region on chromosome 6.
2. Genes for 28S and 5.8S ribosomal RNAs appear in a 44-kb tandem repeat unit of 150—200 copies. 5S RNA genes also

appear in tandem arrays containing 200—300 genes, the largest of which is on chromosome 1.

3. Small nucleolar RNAs include two families of molecules that cleave and process ribosomal RNAs.
4. Spliceosomal snRNAs, including the U1, U2, U4, U5, and U6 snRNAs, many of which appear in clusters of tandem repeats of nearly-identical sequences, or inverted repeats.

2.7 SINGLE-NUCLEOTIDE POLYMORPHISMS (SNPS)

A single-nucleotide polymorphism or SNP (pronounced 'snip') is a genetic variation between individuals, limited to a single base pair which can be substituted, inserted or deleted. Sickle-cell anaemia is an example of a disease caused by a specific SNP: an A → T mutation in the β-globin gene changes a Glu → Val, creating a sticky surface on the haemoglobin molecule that leads to polymerization of the deoxy form. SNPs are distributed throughout the genome, occurring on the average every 2000 base pairs. Although they arose by mutation, many positions containing SNPs have low mutation rates, and provide stable markers for mapping genes.

A consortium of four major academic genome centres and 11 private companies is creating a high-quality, high-density human SNP map, with a commitment to open public access. A recently published version contains 1.42 million SNPs. Not all SNPs are linked to diseases. Many are not within functional regions (although the density of SNPs is higher than the average in regions containing genes). Some SNPs that occur within exons are mutations to synonymous codons, or cause substitutions that do not significantly affect protein function.

Other types of SNPs can cause more than local perturbation to a protein: (1) A mutation from a sense codon to a stop codon, or *vice versa,* will cause either premature truncation of protein synthesis or 'readthrough.' (2) A deletion or insertion will cause a phase shift in translation. The A, B and O alleles of the genes for blood groups illustrate these possibilities. A and B alleles differ by four SNP substitutions. They code for related proteins that add different saccharide units to an antigen on the surface of red blood cells.

Allele	*Sequence*	*Saccharide*
A	...gctggtgacccctt...	N-acetylgalactosamine
B	...gctcgtcaccgcta...	galactose
O	...cgtggt-acccctt...	—

The O allele has undergone a mutation causing a phase shift, and produces no active enzyme. The red blood cells of type O individuals contain neither the A nor the B antigen. This is why people with type O blood are universal donors in blood transfusions. The loss of activity of the protein does not seem to carry any adverse consequences. Indeed, individuals of blood types Band O have greater resistance to smallpox. Strong correlation of a disease with a specific SNP is advantageous in clinical work, because it is relatively easy to test for affected people or carriers. But if a disease arises from dysfunction of a specific protein, there ought to be many sites of mutations that could cause inactivation.

A particular site may predominate if (1) all bearers of the gene are descendants of a single individual in whom the mutation occurred, and/or (2) the disease results from a *gain* rather than loss of a specific property, such as in the ability of sickle-cell haemoglobin to polymerize, and/or (3) the mutation rate at a particular site is unusually high, as in the Gly380→ Arg mutation in the fibroblast growth receptor gene FGFR3, associated with achondroplasia (a syndrome including short stature). In contrast, many independent mutations have been detected in the BRCA1 and BRCA2 genes, loci associated with increased disposition to early-onset breast and ovarian cancer.

The normal gene products function as tumour suppressors. Insertion or deletion mutants causing phase shifts generally produce a missing or inactive protein. But it cannot be deduced a priori whether a novel substitution mutant in BRCA1 or BRCA2 confers increased risk or not. Treatments of diseases caused by defective or absent proteins include:

1. *Providing normal protein*. We have mentioned insulin for diabetes, and Factor VIII in the most common type of haemophilia. Another example is the administration of human growth hormone in patients with an absence or severe reduction in normal levels. Use of recombinant proteins eliminates the risk of transmission of AIDS through blood transfusions or of Creutzfeldt-Jakob disease from growth hormone isolated from crude pituitary extracts.
2. *Life-style adjustments that make the function unnecessary*. Phenylketonuria (PKU) is a genetic disease caused by deficiency in phenylalanine hydroxylase, the enzyme that converts phenylalanine to tyrosine. Accumulation of high levels of

phenylalanine causes developmental defects, including mental retardation. The symptoms can be avoided by a phenylalanine-free diet. Screening of newborns for high blood phenylalanine levels is legally required in the United States and many other countries.

3. Gene therapy to replace absent proteins is an active field of research.

Other clinical applications of SNPs reflect correlations between genotype and reaction to therapy (pharmacogenomics). For example, an SNP in the gene for N-acetyl transferase NAT—2 is correlated with peripheral neuropathy—weakness, numbness and pain in the arms, legs, hands or feet—as a side effect of treatment with isoniazid (isonicotinic acid hydrazide), a common treatment for tuberculosis. Patients with this SNP are given alternative treatment.

2.8 GENETIC DIVERSITY IN ANTHROPOLOGY

SNP data are of great utility in anthropology, giving clues to historical variations in population size, and migration patterns. Degrees of genetic diversity are interpretable in terms of the size of the founding population. Founders are the original set of individuals from whom an entire population is descended. Founders can be either original colonists, such as the Polynesians who first settled in New Zealand, or merely the survivors of a near-extinction. Cheetahs show the effects of a population bottleneck, estimated to have occurred 10 000 years ago. All living cheetahs are as closely related to one another as siblings.

Extrapolations of mitochondrial DNA variation in contemporary humans suggest a single maternal ancestor who lived 140 000–200 000 years ago. Calling her Eve suggests that she was the first woman. But fossil evidence for humans goes back much longer. Mitochondrial Eve was in fact the founder of a surviving population following a near-extinction. Population-specific SNPs are informative about migrations. Mitochondrial sequences provide information about female ancestors and Y chromosome sequences provide information about male ancestors. For example, it has been suggested that the population of Iceland—first inhabited about 1100 years ago—is descended from Scandinavian males, and from females from both Scandinavia and the British Isles. Medieval Icelandic writings refer to raids on settlements in the British Isles.

A fascinating relationship between human DNA sequences and language families has been investigated by L.L. Cavalli-Sforza and colleagues. These studies have proved useful in working out interrelationships among American Indian languages. They confirm that the Basques, known to be a linguistically isolated population, have been genetically isolated also. In the discovery of isolated populations, anthropological genetics provides data useful in medicine, for mapping disease genes is easier if the background variation is low.

Genetically isolated populations in Europe include, in addition to the Basques, the Finns, Icelanders, Welsh, and Lapps. In Iceland, good genealogical and medical records are available. In a controversial move, the government of Iceland passed a law in 1998 authorizing the creation of a database of the medical records, family history and genetic sequences of its 275 000 citizens. A company, Decode Genetics, will collaborate with the worldwide pharmaceutical industry in the analysis and application of these resources.

2.8.1 Genetic Diversity and Personal Identification

Variations in our DNA sequences give us individual fingerprints, useful for identification and for establishment of relationships, including but not limited to questions of paternity. The use of DNA analysis as evidence in criminal trials is now well-established.

Genetic fingerprinting techniques were originally based on patterns of VNTRs, but have been extended to include analysis of other features including mitochondrial DNA sequences. For most of us, all our mitochondria are genetically identical, a condition called homoplasmy. However, some individuals contain mitochondria with different DNA sequences; this is heteroplasmy. Such sequence variation in a disease gene can complicate the observed inheritance pattern of the disease. The most famous case of heteroplasmy involved Tsar Nicholas II of Russia.

After the revolution the Tsar and his family were taken to exile in Yekaterinburg in Central Russia. During the night of July 1918, the Tsar, Tsarina Alexandra, at least three of their five children, plus their physician and three servants who had accompanied the family, were killed, and their bodies buried in a secret grave. When the remains were rediscovered, assembly of the bones and examination of the dental work suggested—and sequence analysis confirmed—that the remains included an expected family group. The identity of the remains of the Tsarina were proved by matching the mitochondrial DNA sequence with that of a maternal relative, Prince Philip, Chancellor of the University of Cambridge, Duke of Edinburgh, and grandnephew of the Tsarina. However, comparisons of mitochondrial DNA sequences of the putative remains of Nicholas II with those of two maternal relatives revealed a difference at base 16169: the Tsar had a C and the relatives a T. The extreme political and even religious sensitivities mandated that no doubts were tolerable. Further tests showed that the Tsar was heteroplasmic; T was a minor component of his mitochondrial DNA at position 16169. To confirm the identity beyond any reasonable question, the body of Grand Duke Georgij, brother of the Tsar, was exhumed, and was shown to have the same rare heteroplasmy.

2.8.2 Genetic Analysis of Cattle Domestication

Animal resources are an integral and essential aspect of human culture. Analysis of DNA sequences sheds light on their historical development and on the genetic variety characterizing modern breeding populations. Contemporary domestic cattle include those familiar in Western Europe and North America, *Bos taurus;* and the zebu of Africa and India, *Bos indicus*.

The most obvious difference in external appearance is the humpback of zebu. It has been widely believed that the domestication of cattle occurred once, about 8000-10 000 years ago, and that the two species subsequently diverged. Analysis of mitochondrial DNA sequences from European, African and Asian cattle suggest however that (1) all European and African breeds are more closely related to each other than either is to Indian breeds, and (2) the two groups diverged about 200 000 years ago, implying recent independent domestications of different species. The similarity in physical appearance of African and Indian zebu (and other similarities at the molecular level; for instance, VNTR markers in nuclear DNA) must then be attributable to importation of cattle from India to East Africa.

2.9 EVOLUTION OF GENOMES

The availability of complete information about genomic sequences has redirected research. A general challenge in analysis of genomes is to identify 'interesting events'. A background mutation rate in coding sequences is reflected in *synonymous* nucleotide substitutions: changes in codons that do not alter the amino acid. With this as a baseline one can search for instances in which there are significantly higher rates of *non-synonymous* nucleotide substitutions: changes in codons that cause mutations in the corresponding protein.

Given two aligned gene sequences, we can calculate K_a = the number of non-synonymous substitutions, and K_s = the number of synonymous substitutions. (The calculation involves more than simple counting because of the need to estimate and correct for possible multiple changes.) A high ratio of K_d/K_s identifies pairs of sequences apparently showing positive selection, possibly even functional changes. The new field of comparative genomics treats questions that can only now be addressed, such as:

- What *genes* do different phyla share? What genes are unique to different phyla? Do the arrangements of these genes in the genome vary from phylum to phylum?
- What homologous *proteins* do different phyla share? What proteins are unique to different phyla? Does the integration of the activities of these proteins vary from phylum to phylum? Do the mechanisms of control of expression patterns of these proteins vary from phylum to phylum?
- What *biochemical functions* do different phyla share? What biochemical functions are unique to different phyla? Does the integration of these biochemical functions vary from phylum to phylum? If two phyla share a function, and the protein that carries out this function in one phylum has a homologue in the other, does the homologous protein carry out the same function?

The same questions could be asked about different species within each phylum. The human genome may have first claim on our attention. But the ability of different genomes to illuminate one another is of the utmost importance.

M.A. Andrade, C. Ouzounis, C. Sander, J. Tamames, and A. Valencia compared the protein repertoire of species from the three major domains of life: *Haemophilus influenzae* represented the bacteria, *Methanococcus jannaschii* the eukarya, and *Saccharomyces cerevisiae* (yeast) the eukarya. Their classification of protein functions contained as major categories processes involving energy, information, and communication and regulation:

General functional classes

- Energy
  - — Biosynthesis of cofactors, amino acids
  - — Central and intermediary metabolism
  - — Energy metabolism
  - — Fatty acids and phospholipids
  - — Nucleotide biosynthesis
  - — Transport

- Information
  - Replication
  - Transcription
  - Translation
- Communication and regulation
  - Regulatory functions
  - Cell envelope/cell wall
  - Cellular processes

The number of genes in the three species are:

Species	*Number of genes*
Haemophilus influenzae	1680
Methanococcus jannaschii	1735
Saccharomyces cerevisiae	6278

Are there, among these, shared proteins for shared functions? In the category of Energy, proteins are shared across the three domains. In the category of Communication, proteins are unique to each domain. In the categories of Regulation and Information, archaea share some proteins with bacteria and others with eukarya. Analysis of shared functions among all domains of life has led people to ask whether it might be possible to define a *minimal organism*—that is, an organism with the smallest gene complement consistent with independent life based on the central DNA→ RNA→protein dogma (i.e. excluding protein-free life forms based solely on RNA).

The minimal organism must have the ability to reproduce, but not be required to compete in growth and reproductive rate with other organisms. One may reasonably assume a generous nutrient medium, relieving the organism of biosynthetic responsibility, and dispensing with stress-response functions including DNA repair. The smallest known independent organism is *Mycoplasma genitalium*, with 468 predicted protein sequences. In 1996, A.R. Mushegian and E.V. Koonin compared the genomes of *M. genitalium* and *H. influenzae*. (At the time, these were the only completely sequenced bacterial genomes.)

The last common ancestor of these widely diverged bacteria lived about two billion years ago. Of 1703 protein-coding genes of *H. influenzae*, 240 are homologues of proteins in *M. genitalium*. Mushegian and Koonin reasoned that all of these must be essential, but might not be sufficient for autonomous life—for some essential functions might be carried out by unrelated proteins in the two organisms. For instance, the common set of 240 proteins left gaps in essential pathways, which could be filled by adding 22 enzymes from *M. genitalium*. Finally, removing functional redundancy and parasite-specific genes gave a list of 256 genes as the proposed necessary and sufficient minimal set. What is in the proposed minimal genome? Functional classes include:

1. Translation, including protein synthesis
2. DNA replication
3. Recombination and repair—a second function of essential proteins involved in DNA replication.
4. Transcription apparatus
5. Chaperone-like proteins
6. Intermediary metabolism—the glycolytic pathway
7. No nucleotide, amino acid, or fatty acid biosynthesis
8. Protein-export machinery
9. Limited repertoire of metabolite transport proteins

It should be emphasized that the viability of an organism with these proteins has not been proven. Moreover, even if experiments proved that some minimal gene content—the proposed set or some other set—is necessary and sufficient, this does not answer the related question of identifying the gene complement of the common ancestor of *M. genitalium* and *H. influenzae*, or of the earliest cellular forms of life. For only 71% of the proposed set of 256 proteins have recognizable homologues among eukaryotic or archaeal proteins. Nevertheless, identification of functions necessarily common to all forms of life allows us to investigate the extent to which different forms of life accomplish these functions in the same ways. Are similar reactions catalyzed in different species by homologous proteins?

Genome analysis has revealed families of proteins with homologues in archaea, bacteria, and eukarya. The assumption is that these have evolved from an individual ancestral gene through a series of speciation and duplication events, although some may be the effects of horizontal transfer. The challenge is to map common function and common proteins. Several thousand protein families have been identified with homologues in archaea, bacteria and eukarya.

Different species contain different amounts of these common families: In bacteria, the range is from *Aquifex aeolicus*, 83% of the proteins of which have archaeal and eukaryotic homologues, to *Borrelia burgdorferi*, in which only 52% of the proteins have archaeal and eukaryotic homologues. Archaeal genomes have somewhat higher percentages (62-71%) of proteins with bacterial and eukaryotic homologues. But only 35% of the proteins of yeast have bacterial and archaeal homologues. Does the common set of proteins carry out the common set of functions?

Among the proteins of the minimal set identified from *M. genitalium*, only ~30% have homologues in all known genomes. Other essential functions must be carried out by unrelated proteins, or in some cases possibly by unrecognized homologues. The protein families for which homologues carry out common functions in archaea, bacteria and eukarya are enriched in those involved in translation:

Protein functional class	*Number of families appearing in all known genomes*
Translation, including ribosome structure	53
Transcription	4
Replication, recombination, repair	5
Metabolism	9
Cellular processes: (chaperones, secretion, cell division, cell wall biosynthesis)	9

The picture is emerging that evolution has explored the vast potential of proteins to different extents for different types of functions. It has been most conservative in the area of protein synthesis.

2.9.1 Please Pass the Genes: Horizontal Gene Transfer

Learning that *S. griseus* trypsin is more closely related to bovine trypsin than to other microbial proteinases, Brian Hartley commented in 1970 that, '... the bacterium must have been infected by a cow'. It was a clear example of lateral or horizontal gene transfer—a bacterium picking up a gene from the soil in which it was growing, which an organism of another species had deposited there. The classic experiments on pneumococcal transformation by O. Avery, C. MacLeod, and M. McCarthy that identified DNA as the genetic material were an even earlier example. In general, horizontal gene transfer is the acquisition of genetic material by one organism from another, by natural rather than laboratory procedures, through some means other than descent from a parent during replication or mating.

Several mechanisms of horizontal gene transfer are known, including direct uptake, as in the pneumococcal transformation experiments, or via a viral carrier. Analysis of genome sequences has shown that horizontal gene transfer is not a rare event, but has affected most genes in microorganisms. It requires a change in our thinking from ordinary 'clonal' or parental models of heredity. Evidence for horizontal transfer includes (1) discrepancies among evolutionary trees constructed from different genes, and (2) direct sequence comparisons between genes from different species:

1. In *E. coli,* 755 ORFs (a total of 547.8 kb, ~18% of the genome) appear to have entered the genome by horizontal transfer after divergence from the *Salmonella* lineage 100 million years ago.
2. In microbial evolution, horizontal gene transfer is more prevalent among operational genes—those responsible for 'housekeeping' activities such as biosynthesis—than among informational genes—those responsible for organizational activities such as transcription and translation. For example, *Bradyrhizobium japonicum,* a nitrogen-fixing bacterium symbiotic with higher plants, has two glutamine synthetase genes. One is similar to those of its bacterial relatives; the other 50% identical to those of higher plants. Rubisco (ribulose1,5-bisphosphate carboxylase/oxygenase), the enzyme that first fixes carbon dioxide at the entry to the Calvin cycle of photosynthesis, has, been passed around between bacteria, mitochondria and algal plastids, as well as undergoing gene duplication. Many phage genes appearing in the *E. coli* genome provide further examples and point to a mechanism of transfer.

Nor is the phenomenon of horizontal gene transfer limited to prokaryotes. Both eukaryotes and prokaryotes are chimeras. Eukaryotes derive their informational genes primarily from an organism related to *Methanococcus,* and their operational genes primarily from proteobacteria with some contributions from cyanobacteria and methanogens. Almost all informational genes from *Methanococcus* itself are similar to those in yeast. Nor is gene transfer necessarily limited to ancient ancestors. It has been suggested (but is now disputed) that numerous bacterial genes have entered the human genome. Conversely, at least eight human genes appeared in the *M. tuberculosis* genome. The observations hint at the model of a 'global organism', a genetic common market, or even a World Wide DNA Web from which organisms download genes at will! How can this be reconciled with the fact that the discreteness of species has been maintained?

The conventional explanation is that the living world contains ecological 'niches' to which individual species are adapted. It is the discreteness of niches that explains the discreteness of species. But this explanation depends on the stability of normal heredity to maintain the fitness of the species. Why would not the global organism break down the lines of demarcation between species, just as global access to pop culture threatens to break down lines of demarcation between national and ethnic cultural heritages? Perhaps the answer is that it is the informational genes, which appear to be less subject to horizontal transfer, that determines the identity of the species. It is interesting that although evidence for the importance of horizontal gene transfer is overwhelming, it was dismissed for a long time as rare and unimportant.

The source of the intellectual discomfort is clear: Parent-to-child transmission of genes is at the heart of the Darwinian model of biological evolution whereby selection (differential reproduction) of parental phenotypes alters gene frequencies in the next generation. For offspring to gain genes from elsewhere than their parents smacks of Lamarck and other discredited alternatives to the paradigm. The evolutionary tree as an organizing principle of biological relationship is a deeply ingrained concept: scientists display an environmentalist-like fervour in their commitment to trees, even when trees are not an appropriate model of a network of relationships. Perhaps it is well to recall that Darwin knew nothing of genes, and the mechanism that generated the variation on which selection could operate was a mystery to him. Maybe he would have accepted horizontal gene transfer more easily than his followers!

2.9.2 Comparative Genomics of Eukaryotes

A comparison of the genomes of yeast, fly, worm and human revealed 1308 groups of proteins that appear in all four. These form a conserved core of proteins for basic functions, including metabolism, DNA replication and repair, and translation. These proteins are made up of individual protein domains, including single-domain proteins, oligomeric proteins, and modular proteins containing many domains (the biggest, the muscle protein titin, contains 250-300 domains.)

The proteins of the worm and fly are built from a structural repertoire containing about three times as many domains as the proteins of yeast. Human proteins are built from about twice as many as those of the worm and fly. Most of these domains appear also in bacteria and archaea, but some are specific to (probably, invented by) vertebrates. These include proteins that mediate activities unique to vertebrates, such as defence and immunity proteins, and proteins in the nervous system; only one of them is an enzyme, a ribonuclease.

2.8. Distribution of probable homologues of predicted human proteins

Vertebrates only	22%
Vertebrates and other animals	24%
Animals and other eukaryotes	32%
Eukaryotes and prokaryotes	21%
No homologues in animals	1%
Prokaryotes only	1%

To create new proteins, inventing new domains is an unusual event. It is far more common to create different combinations of existing domains in increasingly complex ways. A common mechanism

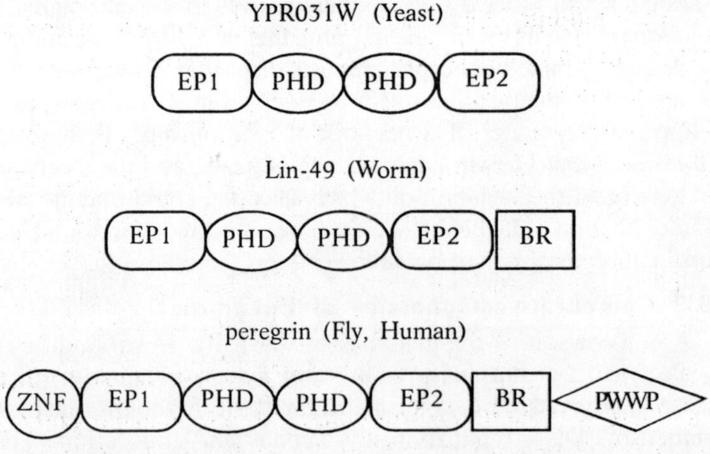

Figure 2.2. Evolution by accretion of domains, of molecules related to perigrin, a human protein that probably functions in transcription regulation.

is by accretion of domains at the ends of modular proteins. This process can occur independently, and take different courses, in different phyla.

Gene duplication followed by divergence is a mechanism for creating protein families. For instance, there are 906 genes + pseudogenes for olfactory receptors in the human genome. These are estimated to bind ~10000 odour molecules. Homologues have been demonstrated in yeast and other fungi (some comparisons *are* odorous), but it is the need of vertebrates for a highly-developed sense of smell that multiplied and specialized the family to such a great extent. Eighty percent of the human olfactory receptor genes are in clusters. Compare the small size of the globin gene cluster, which did not require such great variety.

3

Genomics

Life as we know it is specified by genomes. Every organism possesses a genome that contains the biological information needed to construct and maintain a living example of that organism. Most genomes, including those for all cellular lifeforms, are made of DNA (deoxyribonucleic acid) but a few viruses have RNA (ribonucleic acid) genomes. DNA and RNA are polymeric molecules made up of linear, unbranched chains of monomeric subunits called nucleotides. Each nucleotide has three parts: a sugar, a phosphate group, and a base. In DNA, the sugar is 2'-deoxyribose and the bases are adenine (A), cytosine (C), guanine (G) and thymine (T). Nucleotides are linked to one another by phosphodiester bonds to form a DNA polymer, or polynucleotide, which might be several million nucleotides in length. DNA in living cells is double-stranded, two polynucleotides being wound around one another to form the double helix. The double helix is held together by hydrogen bonds between the base components of the nucleotides in the two strands. The base-pairing rules are that A base-pairs with T, and G base-pairs with C. The two DNA molecules in a double helix therefore have complementary sequences.

In an RNA nucleotide the sugar is ribose rather than 2'-deoxyribose, and thymine is replaced by the related base uracil (U). RNA polymers are rarely more than a few thousand nucleotides in length, and RNA in the cell is usually single-stranded, although base pairs might form between different parts of a single molecule.

The biological information contained in a genome is encoded in the nucleotide sequence of its DNA or RNA molecules and is divided into discrete units called genes. The information contained in a gene is read by proteins that attach to the genome at the appropriate positions and initiate a series of biochemical reactions referred to as gene expression. For organisms with DNA genomes, this process was originally looked on as comprising two stages, transcription and translation, the first producing an RNA copy of the gene and the second resulting in synthesis of a protein whose amino acid sequence is determined, via the genetic code, by the nucleotide sequence of the RNA transcript. This is still an accurate description of gene expression in simple organisms such as bacteria, but it gives an incomplete picture of the events involved in conversion of genomic information into functional proteins in higher organisms. A particular weakness of the transcription-translation inter-pretation is that it can result in attention being drawn away from the key points in the gene expression pathway at which information flow is regulated.

A complete copy of the genome must be made every time a cell divides. DNA replication has to be extremely accurate in order to avoid the introduction of mutations into the genome copies. Some mutations do, however, occur, either as errors in replication or due to the effects of chemical and physical mutagens that directly alter the chemical structure of DNA. DNA repair enzymes correct many of these errors; those that escape the repair processes become permanent features of the lineage descending from the original mutated genome. These events, along with genome rearrangements resulting from recombination, underlie molecular evolution, the driving force behind the evolution of living organisms.

3.1 HUMAN GENOME

Of all the genomes in existence our own is quite naturally the one that interests us the most. We will therefore begin our exploration with an overview of the structure and organization of the human genome. We will then, consider how similar or different the human genome is to the genomes of other organisms.

3.1.1 Physical Structure of Human Genome

The human genome is made up of two distinct components.

1. The *nuclear genome*, which comprises approximately 3000000000 bp (*base pairs*) of DNA. This figure is the same as 3000000kb (*kilobase pairs*) or 3000Mb (*megabase pairs*). The nuclear genome is divided into 24 linear DNA molecules, the shortest 55Mb in length and the longest 250Mb, each

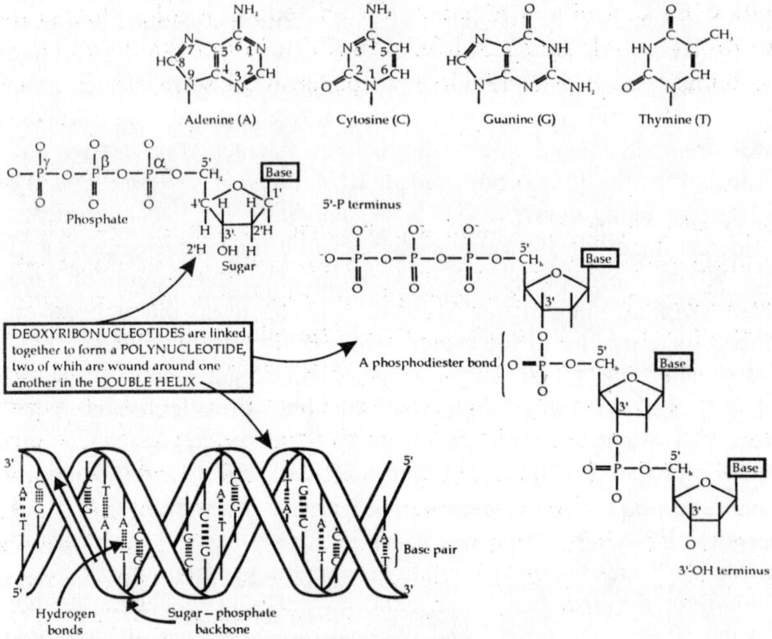

Figure 3.1. The structure of DNA.

contained in a different *chromosome*. These 24 chromosomes consist of 22 *autosomes* and the two *sex chromosomes*, X and Y.

2. The *mitochondrial genome*, a circular DNA molecule of 16569bp, many copies of which are located in the energy-generating organelles called *mitochondria*.

Each of the approximately 10^{13} cells in the adult human body has its own copy or copies of the genome, the only exceptions being those few cell types, such as red blood cells, that lack nuclei in their fully differentiated state. The vast majority of cells are *diploid* and so have two copies of each autosome, plus two sex chromosomes, XX for females or XY for males - 46 chromosomes in all. These are called *somatic cells*, in contrast to *sex cells* or *gametes*, which are *haploid* and have just 23 chromosomes, comprising one of each autosome and one sex chromosome. Both types of cell have about 8000 copies of the mitochondrial genome, 10 or so in each mitochondrion.

Before we go any further we should try to gain some appreciation of the immensity of the human genome. Three billion

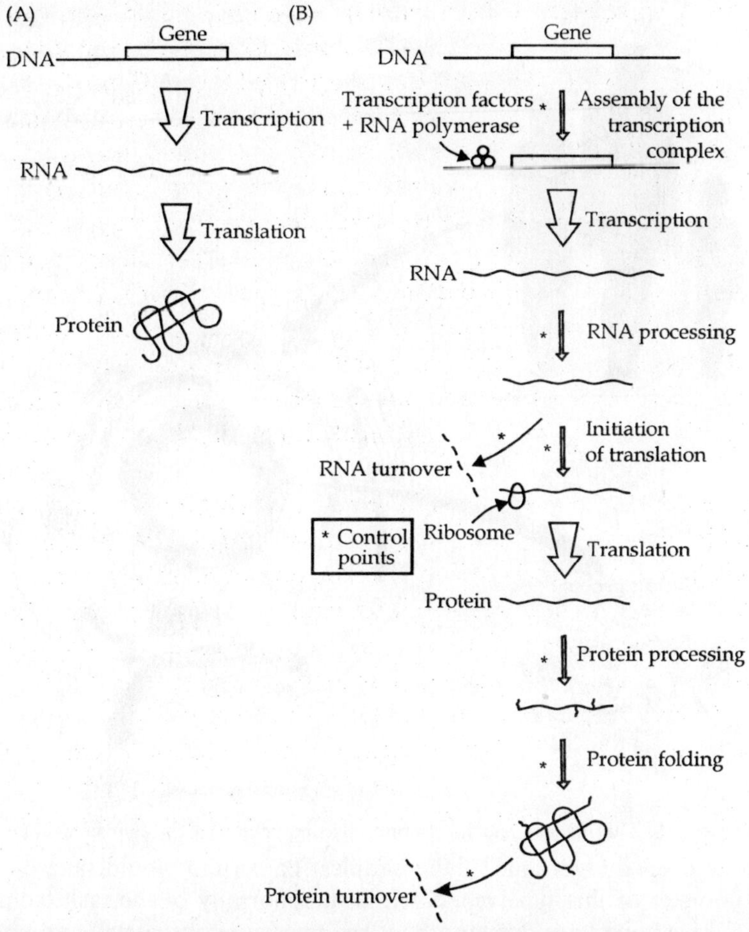

Figure 3.2. Two views of gene expression: (A) shows the traditional depiction of gene expression, summarized as 'DNA makes RNA makes protein', the two steps being called transcription and translation, (B) gives a more accurate outline of the events involved in gene expression in higher organisms.

base pairs is such a large number of nucleotides that it is difficult to grasp the scale that it represents; an analogy is helpful. The typeface used for the text of this book enables approximately 60 nucleotides of DNA sequence to be written in a line 10 cm in length. If printed out in this format, the human genome sequence would stretch for 5000 km, the distance from Montreal to London, Los Angeles to Panama, Tokyo to Calcutta, Cape Town to Addis Ababa, or Auckland to Perth. The sequence would fill about 3000 books the size of this

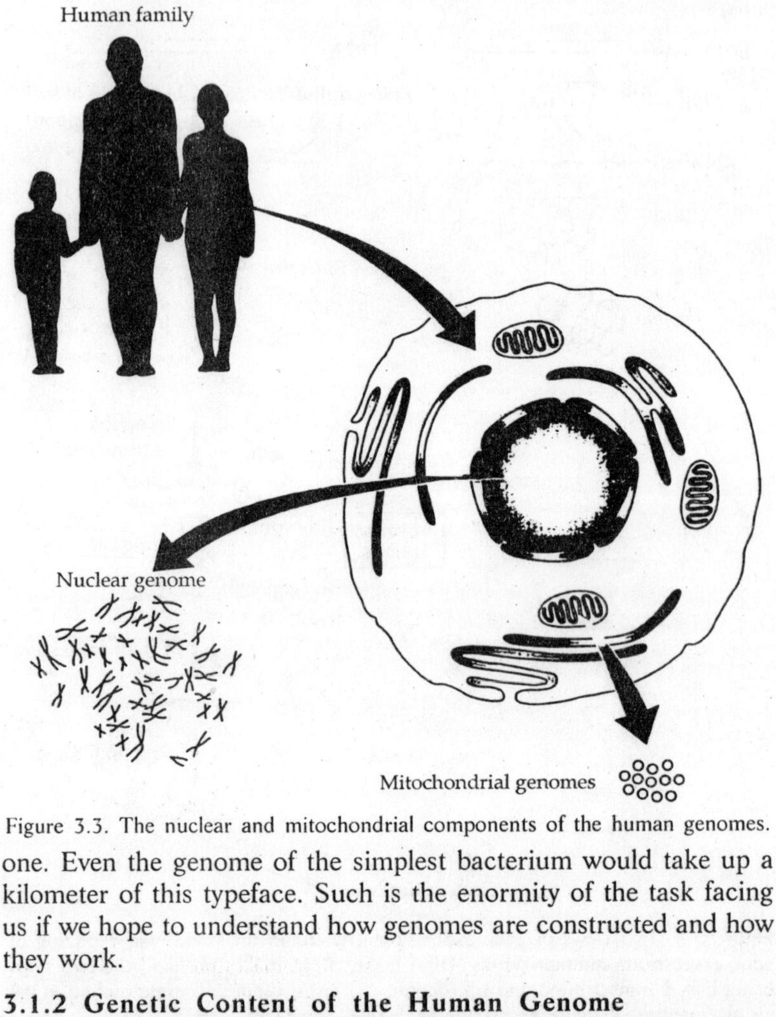

Figure 3.3. The nuclear and mitochondrial components of the human genomes.

one. Even the genome of the simplest bacterium would take up a kilometer of this typeface. Such is the enormity of the task facing us if we hope to understand how genomes are constructed and how they work.

3.1.2 Genetic Content of the Human Genome

Understanding how the human nuclear genome is constructed is one of the goals of the Human Genome Project. First conceived in 1984 and begun in earnest in 1990, the Project aims to complete the sequence of the human genome by 2003. What will the sequence reveal?

The current consensus is that the *human genome* contains approximately 80000 genes, though numbers as low as 50000 and as high as 150000 have been suggested by scientists working on the Human Genome Project. We will also learn that the information contained in these 80000 or so genes takes up only 3% of the nuclear genome. This is illustrated in the figure, which shows the genetic organization of a 50-kb segment of chromosome 7, this segment forming part of the 'human β T-cell receptor locus', a much larger (685kb) region of the genome that specifies proteins involved in the immune response. Our 50-kb segment contains the following genetic features:

3.1.2.1 One gene

This gene is called TRY4 and it codes for trypsinogen, the inactive precursor of the digestive enzyme trypsin. TRY 4 is one of a family of trypsinogen genes present in two clusters at either end of the β T-cell receptor locus. These genes have nothing to do with the immune response, they simply share this part of chromosome 7 with the β T-cell receptor locus.

3.1.2.2 Two gene segments

These are V28 and V29-1 and they code for a part of the β T-cell receptor protein after which the locus is named. V28 and V29-1 are quite unusual as they are not complete genes, only segments of a gene, and before being expressed they must be linked to other gene segments from elsewhere in the locus. This occurs in T lymphocytes and is an example of how a permanent change in the activity of the genome can arise during cellular differentiation. Note that TRY4, V28 and V29-1, like most human genes, are *discontinuous*, being made up of *exons*, containing proteincoding information, separated by noncoding *introns*. The exons, when added together, make up a total of 1414 bp, or 2.8% of the 50-kb segment. The coding capacity of this segment is therefore fairly typical of the genome as a whole.

3.1.2.3 One pseudogene

A *pseudogene* is a nonfunctional copy of a gene, usually one that has mutated so that its biological information has become unreadable. This particular pseudogene is called TRY5 and it is closely related to the functional members of the trypsinogen gene family.

3.1.2.4 Fifty-two genome-wide repeat sequences

These are sequences that recur at many places in the genome. There are four main types of genome-wide repeat, called *LINEs* (long interspersed nuclear elements), *SINEs* (short interspersed nuclear elements), *LTR* (long terminal repeat) elements and *DNA transposons*, and examples of each type are seen in this segment, together making up 39.1% of the sequence.

3.1.2.5 Two microsatellites

These are sequences in which a short motif is repeated in tandem. One of the microsatellites seen here has the motif GA repeated sixteen times, giving the sequence

5'-GAGAGAGAGAGAGAGAGAGAGAGAGAGAGAGA-3'

3'-CTCTCTCTCTCTCTCTCTCTCTCTCTCTCTCT-5'

The second microsatellite comprises six repeats of TATT. Many microsatellites are polymorphic, the number of repeats being variable in different individuals. Microsatellites are useful marker points in the genome.

Finally, approximately 50% of our 50-kb segment of the human genome is made up of stretches of nongenic, nonrepetitive, single-copy DNA of no known function or significance.

No short segment can be considered truly representative of the human genome as a whole. In some respects the 50-kb sequence is probably fairly atypical, but it illustrates the range of genetic features contained in the human genome. The next question that we will address is how similar the human genome is to the genomes of other organisms.

3.2 GENOMES OF OTHER ORGANISMS

Biologists divide the living world into two types of organism. *Eukaryotes* – whose cells contain membrane-bound compartments, including a nucleus and organelles such as mitochondria and, in the case of plant cells, chloroplasts. Eukaryotes include animals, plants, fungi and protozoa. *Prokaryotes* – whose cells lack extensive internal compartments. There are two very different groups of prokaryotes, distinguished from one another by characteristic genetic and biochemical features:

1. The *bacteria*, which include most of the commonly encountered prokaryotes such as the gram-negatives (*e.g. Escherichia coli*), the grampositives (*e.g. Bacillus subtilis*), the cyanobacteria (*e.g. Anabaena*) and many more;
2. The *archaea*, which are less well-studied, and have mostly been found in extreme environments such as hot springs, brine pools and anaerobic lake bottoms.

Eukaryotes and prokaryotes have quite different types of genome and we must therefore consider them separately.

3.2.1 Genomes of Eukaryotes

Humans are fairly typical eukaryotes and the human genome is in many respects a good model for eukaryotic genomes in general. All of the eukaryotic nuclear genomes that have been studied are, like the human version, divided into two or more linear DNA molecules, each contained in a different chromosome, and all eukaryotes also possess smaller, usually circular, mitochondrial genomes. The only general eukaryotic feature not illustrated by the human genome is the presence in plants and other photosynthetic organisms of a third genome, located in the chloroplasts. The two organellar genomes - mitochondrial and chloroplast - will be examined in our detailed survey of genome organization.

Although the basic physical structures of all eukaryotic nuclear genomes are similar, one important feature is very different in different organisms. This is genome size, the smallest eukaryotic genomes being less than 10 Mb in length, and the largest over 100,000 Mb. This size range coincides to a certain extent with the complexity of the organism, the simplest eukaryotes such as fungi having the smallest genomes, and higher eukaryotes such as

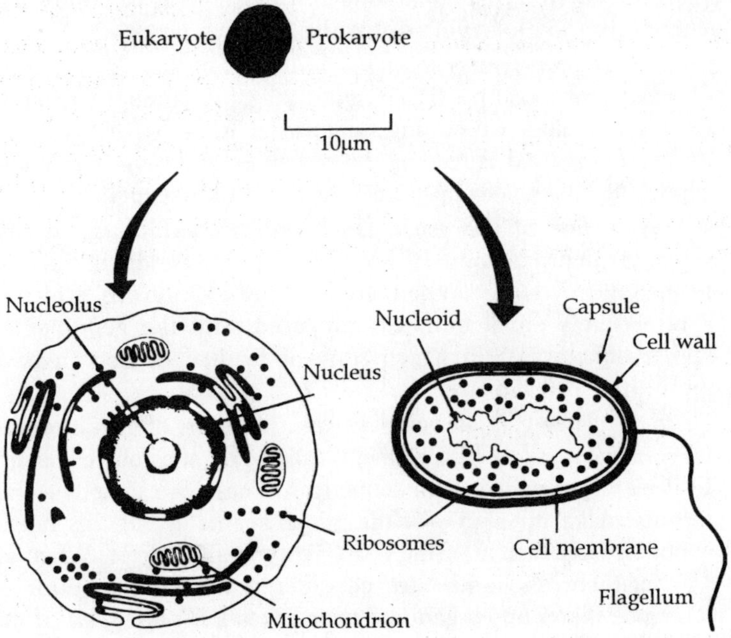

Figure 3.4. Cells of eukaryotes (left) and prokaryotes (right)

vertebrates and flowering plants having the largest ones. This makes sense as one would expect the complexity of an organism to be related to the number of genes in its genome - higher eukaryotes need larger genomes to accom-modate the extra genes. However, the correlation is not precise: if it was, then the nuclear genome of the yeast *Saccharomyces cerevisiae,* which at 12 Mb is 0.004 times the size of the human nuclear genome, would be expected to contain 0.004 × 80000 genes, which is just 320. In fact the S. *cerevisiae* genome contains about 6000 genes.

For many years the lack of precise correlation between the complexity of an organism and the size of its genome was looked on as a bit of a puzzle, the so-called C-value paradox. In fact the answer is quite simple: space is saved in the genomes of less complex organisms because the genes are more closely packed together. The S. *cerevisiae* genome, the sequence of which was completed in 1996, illustrates this point, as we can see from the top two parts of, where the 50-kb segment of the human genome that we looked at earlier in the chapter is compared with a 50-kb segment of the yeast genome. The yeast genome segment, which comes from chromosome III (the first eukaryotic chromosome to be completely sequenced), has the following distinctive features:

1. *It contains more genes than the human segment.* This region of yeast chromosome III contains 26 genes thought to code for proteins and two coding for transfer RNAs (tRNAs), short molecules involved in reading the genetic code during translation. These 28 genes take up 66.4% of the 50-kb sequence.
2. *Relatively few of the yeast genes are discontinuous.* In this segment of chromosome III none of the genes are discontinuous. In the entire yeast genome there are only 239 introns, which is a remarkably small number compared with the genomes of higher eukaryotes, in which some individual genes can have more than 100 introns.
3. *There are fewer genome-wide repeats.* This part of chromosome III contains a single LTR element, called *Ty2,* and four truncated LTR elements called delta sequences. These five genome-wide repeats make up 13.5% of the 50-kb segment, but this figure is not entirely typical of the yeast genome as a whole. When all 16 yeast chromosomes are considered the total amount of sequence taken up by genome-wide repeats is only 3.4% of the total.

The picture that emerges is that the genetic organization of the yeast genome is much more economical than that of the human version. The genes themselves are more compact, having fewer introns, and the spaces between the genes are relatively short with much less space taken up by genome-wide repeats and other noncoding sequences. We will see that the more complex eukaryotes have less compact genomes holds when other species are examined. We will also see that the genome-wide repeats appear to play an intriguing role in dictating the compactness or otherwise of a genome, a general rule being that the smaller genomes lack extensive repetition, but the larger ones display a proliferation in the number of repeat sequences that are present. This is strikingly illustrated by the maize genome, which at 5000 Mb is larger than the human genome but still relatively small for a flowering plant. Only a few limited regions of the maize genome have been sequenced, but some remarkable results have been obtained, revealing a genome dominated by repetitive elements. A 50-kb segment of this genome, either side of one member of a family of genes coding for the alcohol dehydrogenase enzymes. This is the only gene in this 50-kb region, though there is a second one, of unknown function, approximately 100 kb beyond the right-hand end of the sequence shown here. Instead of genes, the dominant feature of this genome segment is the genome-wide repeats. The majority of these are of the LTR element type, which comprise virtually all of the noncoding part of the segment, and on their own are estimated to make up approximately 50% of the maize genome.

These examples show that some of the features of eukaryotic genomes are surprisingly unusual. We will examine what little is known about the evolutionary events that led to these diverse designs.

3.2.2 Genomes of Prokaryotes

Prokaryotic genomes are very different from eukaryotic ones. There is some overlap between the largest prokaryotic and smallest eukaryotic genomes, but on the whole prokaryotic genomes are much smaller. The E. *coli* genome, for example, is just 4639 kb, two-fifths the size of the yeast genome. The physical organizations are also different. In prokaryotes, most if not all of the genome is contained in a single DNA molecule, and this molecule is circular rather than linear. In addition to their single 'chromosome', prokaryotes may also have additional genes on independent, smaller, circular or linear DNA molecules called *plasmids*. Genes carried by plasmids are useful, coding for properties such as antibiotic resistance or the ability to utilize complex compounds such as toluene as a carbon source; but plasmids appear to be dispensable – a prokaryote can exist quite effectively without them. For this reason, a bacterial or archaeal 'genome' is usually defined as just the main DNA molecule, with plasmids looked on as ancillary components, not parts of the genome itself.

As well as having smaller genomes, prokaryotes generally have fewer genes than eukaryotes. *E. coli* has just 4397. After our discussion regarding eukaryotic gene organization, it will probably come as no surprise to learn that prokaryotic genomes are even more compact than those of lower eukaryotes. A 50-kb segment of the *E. coli* genome. It is immediately obvious that there are more genes and less space between them, with 43 genes taking up 85.9% of the segment. Some genes have virtually no space between them: *thrA* and *thrB,* for example, are separated by a single nucleotide, and *thrC* begins at the nucleotide immediately following the last nucleotide of *thrB*. These three genes are an example of an *operon*, a group of genes involved in a single biochemical pathway (in this case, synthesis of the amino acid threonine) and expressed in conjunction with one another. Operons have been used as model systems for understanding how gene expression is regulated.

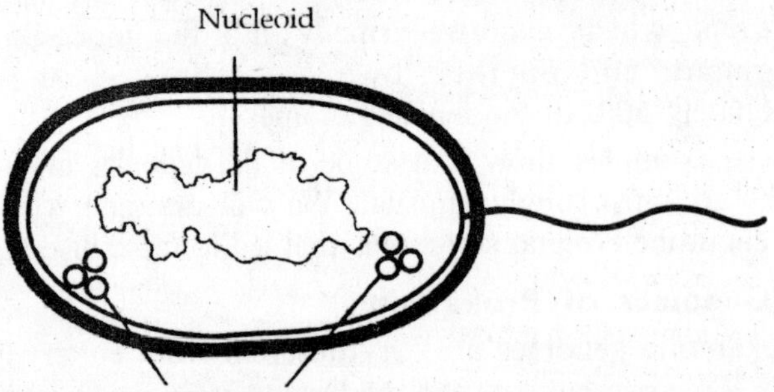

Figure 3.5. Plasmid are small circular DNA molecules that are found inside some prokaryotic cells.

First, there are no introns in the genes present in this segment of the *E. coli* genome. In fact *E. coli* has no discontinuous genes at all and it is generally believed that this type of gene structure is absent in prokaryotes with a few exceptions, mainly among the archaea. The second feature is the infrequency of repetitive sequences. Prokaryotic genomes do not have anything equivalent to the high-copy-number, genome-wide repeat families found in eukaryotic genomes. These are examples of *transposable elements*, sequences that have the ability to move around the genome and to transfer from one organism to another, even sometimes between two different species. The positions of the ISI and ISI elements shown in Figure, therefore refer only to the particular *E. coli* strain from which this sequence was obtained: if a different strain is examined then the IS sequences could well be in different positions or might be entirely absent from the genome.

3.3 WHY ARE GENOME PROJECTS IMPORTANT?

As we begin the new millennium, the major goal of molecular biology is to obtain the complete sequences of as many genomes as possible. Considerable effort is being devoted not only to the Human Genome Project, but also to equivalent projects aimed at the genomes of a wide range of other organisms. Why all this activity devoted to genome sequences? There are many reasons.

First, our current perception is that genome sequences are the key to the continued development of not only molecular biology and genetics, but also those areas of biochemistry, cell biology and physiology now described as the *molecular life sciences*. A catalog containing a description of the sequence of every gene in a genome is immensely valuable, even if at first the functions of many of the genes are unknown. Not only will the catalog contain the sequences of the coding parts of every gene, it will also include the regulatory regions for these genes. The genome sequence therefore opens the way to a comprehensive description of the molecular activities of living cells and the ways in which these activities are controlled.

Gene catalogs also aid the isolation and utilization of important genes, such as those human genes responsible for inherited disease, or bacterial genes whose protein products have industrial value. These genes can be isolated from a genome even if the complete genome sequence is not known, but the process is time-consuming and costly, and a different project has to be devised for each gene that is sought. It will be very much easier to obtain a copy of the desired gene if the sequence of its genome is already known, so that the gene can simply be withdrawn from the catalog.

Genome sequences will have additional benefits that at present can only be guessed at. We have seen that the human genome, in common with the genomes of all higher eukaryotes, contains extensive amounts of noncoding DNA. We assume that most of the noncoding

DNA has no function, but perhaps this is because we do not know enough about it. Could the noncoding DNA have a role, but one that at present is too subtle for us to grasp? The first step in addressing this possibility is to obtain a complete description of the organization of the noncoding DNA in different genomes, so that common features, which might indicate a role for some or all of these sequences, can be identified.

There is one final reason for genome projects. The work stretches current technology to its limits. Genome sequencing therefore represents the frontier of molecular biology, territory that was inaccessible just a few years ago and which still demands innovative approaches and a lot of sheer hard work. Scientists have always striven to achieve the almost impossible, and the motivation for many molecular biologists involved in genome projects is, quite simply, the challenge of the unknown.

4

Detecting Sequence

The ultimate objective of a genome project is the complete DNA sequence for the organism being studied, ideally integrated with the genetic and/or physical maps of the genome so that genes and other interesting features can be located within the DNA sequence. This chapter describes the techniques and research strategies that are used during the sequencing phase of a genome project, when this ultimate objective is being directly addressed. Techniques for sequencing DNA are clearly of central importance in this context and we will begin the chapter with a detailed examination of sequencing methodology. This methodology is, however, of little value unless the short sequences that result from individual sequencing experiments can be linked together in the correct order to give the master sequences of the chromosomes that make up the genome. The middle part of this chapter describes the strategies used to ensure that the master sequences are assembled correctly. Finally, we will consider how the sequencing phase of the Human Genome Project is being conducted.

4.1 METHODOLOGY FOR DNA SEQUENCING

Rapid and efficient procedures for DNA sequencing were first devised in the mid-1970s. Two different procedures were published at almost the same time:

1. *The chain termination method,* in which the sequence of a single-stranded DNA molecule is determined by enzymatic synthesis of complementary polynucleotide chains, these chains terminating at specific nucleotide positions.

2. The chemical degradation method, in which the sequence of a double-stranded DNA molecule is determined by treatment with chemicals that cut the molecule at specific nucleotide positions.

Both methods were equally popular to begin with but the chain termination procedure has gained ascendancy in recent years, especially for genome sequencing. This is partly because the chemicals used in the chemical degradation method are toxic and therefore hazardous to the health of the researchers doing the sequencing experiments, but mainly because it has been easier to automate chain termination sequencing. As we will see later in this chapter, a genome project involves a huge number of individual sequencing experiments and it would take many years to carry all these out by hand. Automated sequencing techniques are therefore essential if the project is to be completed in a reasonable timespan.

4.1.1 Chain Termination DNA Sequencing

Chain termination DNA sequencing is based on the principle that single-stranded DNA molecules that differ in length by just a single nucleotide can be separated from one another by polyacrylamide gel electrophoresis. This means that it is possible to resolve a family of molecules, representing all lengths from 10 to 1500 nucleotides, into a series of bands.

4.1.2 Chain Termination Sequencing in Outline

The starting material for a chain termination sequencing experiment is a preparation of identical single-stranded DNA molecules. The first step is to *anneal* a short oligonucleotide to the same position on each molecule, this oligonucleotide subsequently acting as the *primer* for synthesis of a new DNA strand, complementary to the *template*. The strand synthesis reaction, which is catalyzed by a DNA polymerase enzyme and requires the four deoxyribonucleotide triphosphates (dATP, dCTP, dGTP and dTTP) as substrates, would normally continue until several thousand nucleotides had been polymerized. This does not occur in a chain termination sequencing experiment because, as well as the four dNTPs, a small amount of a *dideoxynucleotide* (e.g. ddATP) is added to the reaction. The polymerase enzyme does not discriminate between dNTPs and ddNTPs, so the dideoxynucleotide can be incorporated into the growing chain, but it then blocks further elongation because it lacks the 3'-hydroxyl group needed to form a connection with the next nucleotide.

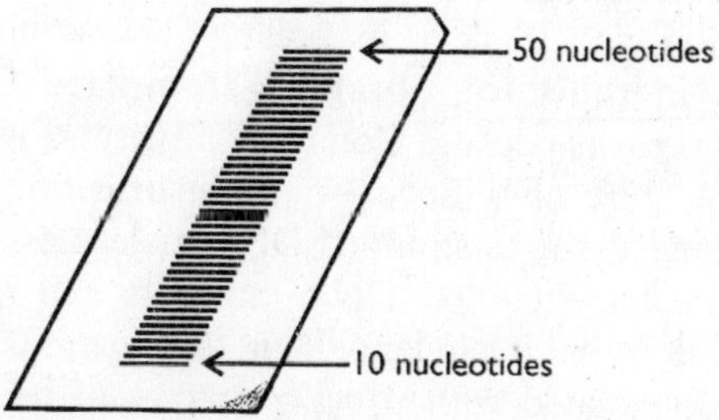

Figure 4.1. Polyacrylamide gel electrophoresis can resolve single-stranded DNA molecules that differ in length by just one nucleotide.

If ddATP is present, chain termination occurs at positions opposite thymidines in the template DNA. Because dATP is also present the strand synthesis does not always terminate at the first T in the template, in fact it may continue until several hundred nucleotides have been polymerized before eventually a ddATP is incorporated. The result is therefore a set of new chains, all of different lengths, but each ending in ddATP. Now the polyacrylamide gel comes into play. The family of molecules generated in the presence of ddATP are loaded into one lane of the gel, and the families generated with ddCTP, ddGTP and ddTTP loaded into the three adjacent lanes. After electrophoresis, the DNA sequence can be read directly from the positions of the bands in the gel. The band that has moved the furthest represents the smallest piece of DNA, this being the strand that terminated by incorporation of a ddNTP at the first position in the template. In the example this band lies in the 'G' lane (*i.e.* the lane containing the molecules terminated with ddGTP), so the first nucleotide in the sequence is 'G'. The next band, corresponding to the molecule that is one nucleotide longer than the first, is in the 'A' lane, so the second nucleotide is 'A' and the sequence so far is 'GA'. Continuing up through the gel we see that the next band also lies in the *'A'* lane (sequence *GAA*), then we move to the *'T'* lane (*GAAT*), and so on. The sequence can be continued up to the region of the gel where individual bands are not separated.

4.1.2.1 Chain termination sequencing requires a single-stranded DNA template

The template for a chain termination experiment is a single-stranded version of the DNA molecule to be sequenced. There are several ways in which this can be obtained:

1. *The DNA can be cloned in a plasmid vector*. The resulting DNA will be double-stranded so cannot be used directly in sequencing.

Instead, it must be converted into single-stranded DNA by denaturation with alkali or by boiling. This is probably the most common method for obtaining template DNA for DNA sequencing, partly because cloning in a plasmid vector is a routine technique, but also because denaturation results in two complementary, single-stranded DNA molecules. This means that both ends of the cloned DNA molecule can be sequenced. A shortcoming is that it can be difficult to prepare plasmid DNA that is not contaminated with small quantities of bacterial DNA and RNA which can act as spurious templates or primers in the DNA sequencing experiment.

2. *The DNA can be cloned in an M13 vector*. Vectors based on M13 bacteriophage are designed specifically for the production of single-stranded templates for DNA sequencing. M13 bacteriophage has a single-stranded DNA genome which, after infection of *Escherichia coli* bacteria, is converted into a doublestranded *replicative form*. The replicative form is copied until over 100 molecules are present in the cell, and when the cell divides the copy number in the new cells is maintained by further replication. At the same time, the infected cells continually secrete new M13 phage particles, approximately 1000 per generation, these phages containing the single-stranded version of the genome. M13 vectors are double-stranded DNA molecules based on the replicative form of the M13 genome. They can be manipulated in exactly the same way as a plasmid cloning vector. The difference is that cells that have been

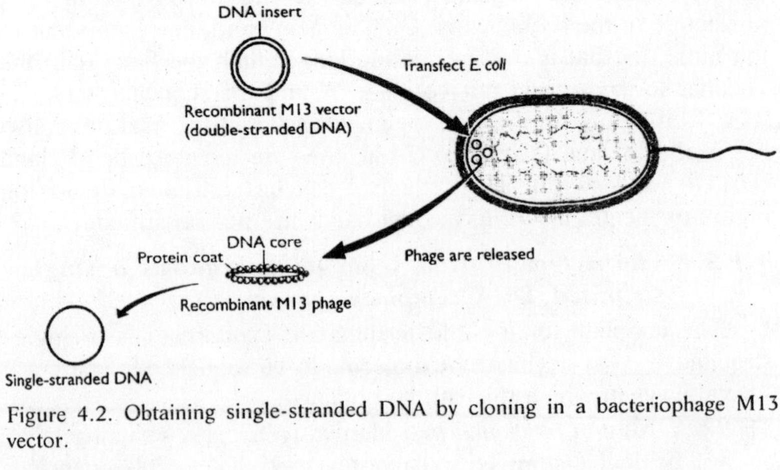

Figure 4.2. Obtaining single-stranded DNA by cloning in a bacteriophage M13 vector.

transfected with a recombinant M13 vector secrete phage particles containing single-stranded DNA, this DNA comprising the vector molecule plus any additional DNA that has been ligated into it. The phages therefore provide the template DNA for chain termination sequencing. Their one disadvantage is that DNA fragments longer than about 3 kb suffer deletions and rearrangements when cloned in an M13 vector, so the system can only be used with short pieces of DNA.

3. *The DNA can be cloned in a phagemid*. This is a plasmid cloning vector that contains, as well as its plasmid origin of replication, the origin from M13 or another phage with a single-stranded DNA genome. If an E. *coli* cell contains both a phagemid and a *helper phage*, the latter carrying genes for the phage replication enzymes and coat proteins, then the phage origin of the phagemid becomes activated, resulting in synthesis

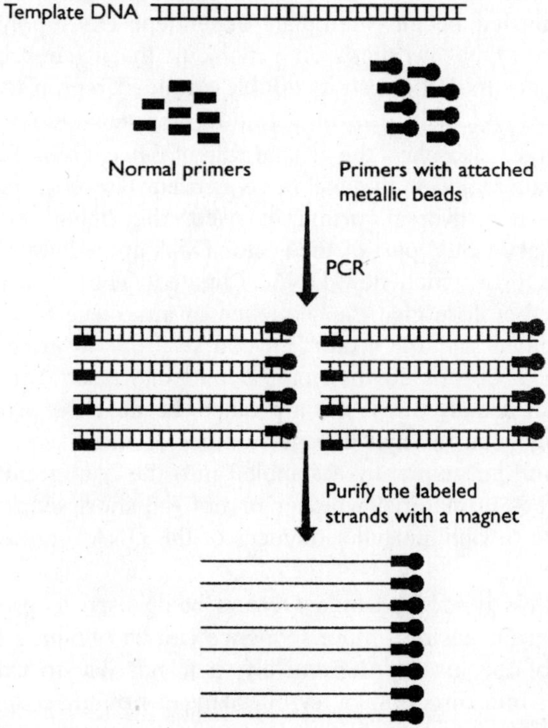

Figure 4.3. One way of using PCR to prepare template DNA for chain termination sequencing.

of phages containing the single-stranded version of the phagemid. The double-stranded plasmid DNA can therefore be converted into single-stranded template DNA for DNA sequencing. This system avoids the instabilities of M13 cloning and can be used with fragments up to lOkb or more.

4. *PCR can be used to generate single-stranded DNA*. There are various methods, the most effective being to modify one of the two primers so that DNA strands synthesized from this primer are easily purified. One possibility is to attach small metallic beads to the primer and then use a magnetic device to purify the resulting strands.

4.1.2.2 Primer determines the region of the template DNA that will be sequenced

To begin a chain termination sequencing experiment an oligonucleotide primer is annealed on to the template DNA. The primer is needed because template-dependent DNA polymerases cannot initiate DNA synthesis on a molecule that is entirely single-stranded: there must be a short double-stranded region to provide a 3' end onto which the enzyme can add new nucleotides.

The primer also plays the critical role of determining the region of the template molecule that will be sequenced. For most sequencing experiments a 'universal' primer is used, this being one that is complementary to the part of the vector DNA immediately adjacent to the point into which new DNA is ligated. The same universal primer can therefore give the sequence of any piece of DNA that has been ligated into the vector. Of course if this inserted DNA is longer than 750bp or so then only a part of its sequence will be obtained, but usually this is not a problem because the project as a whole simply requires that a large number of short sequences are generated and subsequently assembled into the contiguous master sequence. It is immaterial whether or not the short sequences are the complete or only partial sequences of the DNA fragments used as templates.

If double-stranded plasmid DNA is being used to provide the template then, if desired, more sequence can be obtained from the other end of the insert. Alternatively, it is possible to extend the sequence in one direction by synthesizing a nonuniversal primer, designed to anneal at a position corresponding to the end of the longest sequence so far, obtained. An experiment with this primer will provide a second short sequence, overlapping the previous one.

4.1.2.3 Thermal cycle sequencing offers an alternative to the traditional methodology

The discovery of thermostable DNA polymerases, which led to the development of PCR has also resulted in new methodologies for chain termination sequencing. In particular, the innovation called thermal cycle sequencing has two advantages over traditional chain termination sequencing. First, it uses double-stranded rather than single-stranded DNA as the starting material. Second, very little template DNA is needed, so the DNA does not have to be cloned before being sequenced.

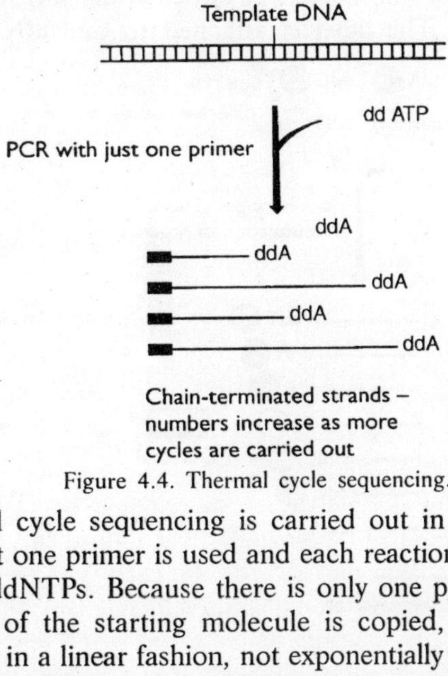

Figure 4.4. Thermal cycle sequencing.

Thermal cycle sequencing is carried out in a similar way to PCR but just one primer is used and each reaction mixture includes one of the ddNTPs. Because there is only one primer only one of the strands of the starting molecule is copied, and the product accumulates in a linear fashion, not exponentially as is the case in a real PCR. The presence of the ddNTP in the reaction mixture causes chain termination, as in the standard methodology, and the family of resulting strands can be analyzed and the sequence read in the normal manner by polyacrylamide gel electrophoresis.

4.1.2.4 Fluorescent primers are the basis of automated sequence reading

The standard chain termination sequencing methodology employs radioactive labels with the banding pattern in the polyacrylamide gel visualized by autoradiography. Usually one of the nucleotides in the sequencing reaction is labeled so that the newly synthesized strands contain radiolabels along their lengths, giving a high detection sensitivity. To ensure- good band resolution, ^{33}P or ^{35}S is generally used, as the emission energies of these isotopes are relatively low, in contrast to ^{32}P, which has a higher emission energy and gives poorer band resolution because of signal scattering.

As earlier we saw how the replacement of radioactive labels by fluorescent ones has given a new dimension to *in situ* hybridization techniques. Fluorolabeling has been equally important in the development of sequencing methodology, in particular because the detection system for fluorolabels has opened the way to automated sequence reading. The label is attached to the ddNTPs, with a

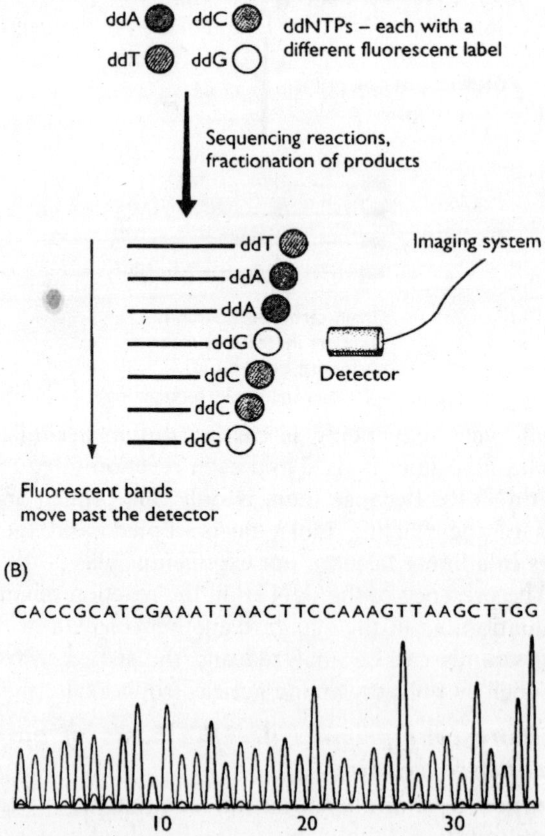

Figure 4.5. Automated DNA sequencing with fluorescently labeled dideoxy-nucleotides.

different fluorolabel being used for each ddNTP. Chains terminated with A are therefore labeled with one fluorophore, chains terminated with C are labeled with a second fluorophore and so on. Now it is possible to carry out the four sequencing reactions - for A, C, G and T - in a single tube and to load all four families of molecules into just one lane of the polyacrylamide gel, because the fluorescent detector can discriminate between the different labels and hence determine if each band represents an A, C, G or T. The sequence can be read directly as the bands pass in front of the detector and either printed out in a form readable by eye or sent straight to a computer for storage. When combined with robotic devices that prepare the sequencing reactions and load the gel, the fluorescent detection system provides a major increase in throughput and avoids errors that might arise when a sequence is read by eye and then entered manually into a computer. It is only by use of these automated techniques that we can hope to generate sequence data rapidly enough to complete a genome project in a reasonable length of time.

4.1.3 Radical Departures from Conventional DNA Sequencing

In spite of the development of automated techniques, conventional DNA sequencing suffers from the limitation that only a few hundred bp of sequence can be determined in a single experiment. In the context of the Human Genome Project, this means that each experiment provides only one five-millionth of the total genome sequence. Attempts are continually being made to modify the technology so that sequence acquisition is more rapid, a recent example being the introduction of new automated sequencers that use capillary separation rather than a polyacrylamide gel. These have 96 channels so 96 sequences can be determined in parallel, and each run takes less than 2 hours to be completed, enabling up to 1000 sequences to be obtained in a single day.

There have also been attempts to make sequence acquisition more rapid by devising new sequencing methodologies. One possibility is pyrosequencing, which does not require electrophoresis or any other fragment separation procedure and so is more rapid than chain termination sequencing. In pyrosequencing, the template is copied in a straightforward manner without added ddNTPs. As the new strand is being made, the order in which the dNTPs are incorporated is detected, so the sequence can be 'read' as the reaction proceeds. The addition

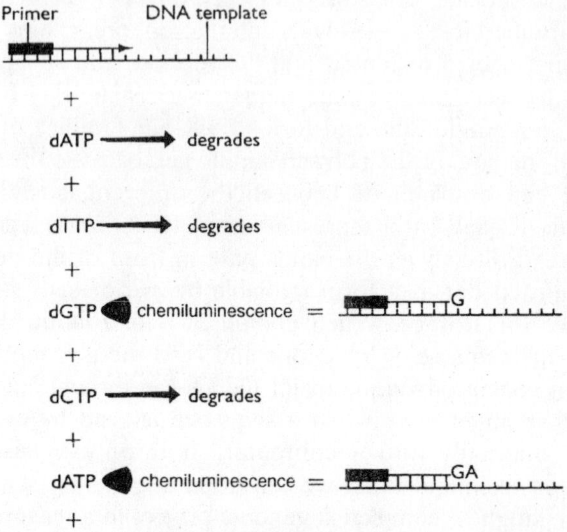

Figure 4.6. Pyrosequencing.

of a nucleotide to the end of the growing strand is detectable because it is accompanied by release of a molecule of pyrophosphate, which can be converted by the enzyme sulfurylase into a flash of chemiluminescence. Of course, if all four dNTPs were added at once then flashes of light would be seen all the time and no useful sequence information would be obtained. Each dNTP is therefore added separately, one after the other, with a nucleotidase enzyme also present in the reaction mixture so that if a dNTP is not incorporated into the polynucleotide then it is rapidly degraded before the next dNTP is added. This procedure makes it possible to follow the order with which the dNTPs are incorporated into the growing strand. The technique sounds complicated, but it simply requires that a repetitive series of additions be made to the reaction mixture, precisely the type of procedure that is easily automated with the possibility of many experiments being carried out in parallel.

A very different approach to DNA sequencing might one day be possible through the use of DNA chips. A chip carrying an array of different oligonucleotides could be used in DNA sequencing by applying the test molecule, the one whose sequence is to be determined, to the array and determining at which positions it hybridizes. Hybridization to an individual oligonucleotide would indicate the presence of that particular oligonucleotide sequence in the test molecule, and comparison of all the oligonucleotides to which hybridization occurs would enable the sequence of the test molecule to be deduced. The problem with this approach is that the maximum length of the molecule that can be sequenced is given by the square root of the number of oligonucleotides in the array, so if every possible 8-mer oligonucleotide (ones containing eight nucleotides) were attached to the chip all 65536 of them - then the maximum length of readable sequence would only be 256 bp. Even if the chip carried all the 1048576 different 10-mer sequences it could still only be used to sequence a 1 kb molecule. To sequence a 1 Mb molecule (this being the sort of advance in sequence capability that is really needed) the chip would have to carryall of the 1×10^2 possible 20-mers. This may sound an outlandish proposition but advances in miniaturization, together with the possibility of electronic rather than visual detection of hybridization, could bring such an array within reach in the future.

4.2 ASSEMBLY OF A CONTIGUOUS DNA SEQUENCE

The next question to address is how the master sequence of a chromosome, possibly several tens of Mb in length, can be assembled from the multitude of short sequences generated by chain termination sequencing. There are three solutions to this problem: the shotgun approach, the clone contig approach, and the directed shotgun approach.

4.2.1 Sequence Assembly by the Shotgun Approach

The straightforward approach to sequence assembly is to build up the master sequence directly from the short sequences obtained from individual sequencing experiments, simply by examining the sequences for overlaps. This is called the shotgun approach. It does not require any prior knowledge of the genome and so can be carried out in the absence of a genetic or physical map.

4.2.1.1 Potential of the shotgun approach was proven by the Haemophilus influenzae sequence

During the early 1990s there was extensive debate about whether the shotgun approach would work in practice, many molecular biologists being of the opinion that the amount of data handling needed to compare all the minisequences and identify overlaps, even with the smallest genomes, would be beyond the capabilities of existing computer systems. These doubts were laid to rest in 1995 when the sequence of the 1830 kb genome of the bacterium *Haemophilus influenzae* was published.

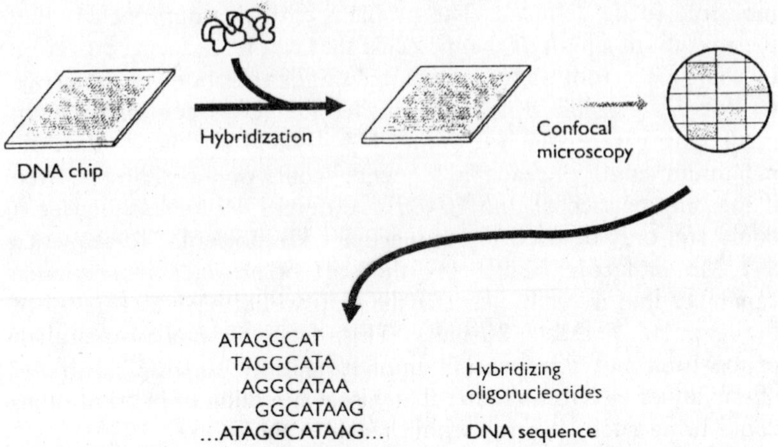

Figure 4.7. A possible way of using chip technology in DNA sequencing.

The *H. influenzae* genome was sequenced entirely by the shotgun approach and without recourse to any genetic or physical map information. The first step was to break the genomic DNA into fragments by sonication, a technique which uses high frequency sound waves to make random cuts in DNA molecules. The fragments were then electrophoresed and those in the range 1.6 to 2.0 kb purified from the agarose gel and ligated into a plasmid vector. From the resulting library, 19687 clones were taken at random and 28643 sequencing experiments carried out, the number of sequencing experiments being greater than the number of plasmids because both ends of some inserts were sequenced. Of these sequencing experiments, 16% were considered to be failures because they resulted in less than 400 bp of sequence. The remaining 24304 sequences gave a total of 11631485 bp, corresponding to six times the length of the H. *influenzae* genome, this amount of redundancy being deemed necessary to ensure complete coverage. Sequence assembly required 30 hours of time on a computer with 512 Mb of RAM, and resulted in 140 lengthy, contiguous sequences, each of these *sequence contigs* representing a different, nonoverlapping portion of the genome.

The next step was to join up pairs of contigs by obtaining sequences from the gaps between them. First, the library was checked to see if there were any clones whose end sequences were located in different contigs. If such a clone could be identified, then additional sequencing of its insert would close the 'sequence gap' between the two contigs. In fact, there were 99 clones in this category, so 99 of the gaps could be closed without too much difficulty.

This left 42 gaps which probably consisted of DNA sequences that were unstable in the cloning vector and therefore not present in the library. To close these 'physical gaps' a second clone library was prepared, this one with a different type of vector. Rather than using another plasmid, in which the uncloned sequences would probably still be unstable, the second library was prepared in a bacteriophage λ vector. This new library was probed with 84 oligonucleotides, one at time, these 84 oligonucleotides having sequences identical to the sequences at the ends of the unlinked contigs. The rationale was that if two oligonucleotides hybridized to the same λ clone then the ends of the contigs from which they were derived must lie within that clone, and sequencing the DNA in the λ clone would therefore close the gap. Twenty three of the 42 physical gaps were dealt with in this way.

A second strategy for gap closure was to use pairs of oligonucleotides, from the set of 84 described above, as primers for PCRs of H. *influenzae* genomic DNA. Some oligonucleotide pairs were selected at random and those spanning a gap identified simply from whether or not they gave a PCR product. Sequencing the resulting PCR products closed the relevant gaps. Other primer pairs were chosen on a more rational basis. For example, oligonucleotides were tested as probes with a Southern blot of H. *influenzae* DNA, and pairs which hybridized to similar sets of restriction fragments identified. The two members of an oligonucleotide pair identified in this way must be contained within the same restriction fragments and so are are likely to lie close together on the genome. This means that the pair of contigs that the oligonucleotides are derived from are adjacent, and the gap between them can be spanned by a PCR using the two oligonucleotides as primers, which will provide the template DNA for gap closure.

4.2.2 Strengths and Limitations of the Shotgun Approach

The demonstration that a small genome can be sequenced relatively rapidly by the shotgun approach led in the late 1990s to a sudden plethora of completed microbial genomes. These projects demonstrated that shotgun sequencing can be set up on a production-line basis, with each team member having his or her individual task in DNA preparation, carrying out the sequencing reactions, or analyzing the data. This strategy resulted in the 580 kb genome of *Mycoplasma genitalium* being sequenced by five people in just 8 weeks and it is now accepted that a year should be ample time to generate the complete sequence of any genome less than about 5Mb, even if nothing is known about the genome before the project begins. The strengths of the shotgun approach are therefore its speed and its ability to work in the absence of a genetic or physical map.

The main weakness is the complexity and limitations of the data analysis needed to identify overlaps between the starting sequences and construct the sequence contigs. Every sequence that is obtained must be compared with every other sequence in order to identify overlaps. Shortcuts can be taken - for example, once a sequence has been attached to a contig it can be taken out of the analysis, reducing the number of comparisons that have to be made with the remaining mini-sequences, but the amount of data analysis is still huge and pushes existing computer systems to their limits. The data analysis is also dependent on there being zero or at worst very little sequence repetition in the genome being studied, as noncontiguous sequences could mistakenly be joined together by 'jumping' from one repeat unit to another. There are many interesting genomes that are small and lack significant numbers of repeated sequences, and the shotgun approach will continue to play an important role at this end of the sequencing project spectrum, but we must look at more structured approaches if we wish to sequence larger genomes.

4.2.3 Sequence Assembly by the Clone Contig Approach

The clone contig approach is the conventional method for obtaining the sequence of a eukaryotic genome and has also been used with those microbial genomes which have previously been mapped by genetic and/or physical means. The clone contig approach accepts that shotgun sequencing is the most efficient method of generating sequence contigs of up to 5Mb, and therefore aims to use partial restriction to break a larger genome down into a library of overlapping, cloned fragments in this size range. Ideally these overlapping fragments are anchored onto a genetic and/or physical map of the genome, so that the sequence data from individual clones can be checked and interpreted by looking for features (e.g. STSs, SSLPs, genes) known to be present in a particular region. To understand the sequence contig approach we must first examine the types of cloning vector available for handling large fragments of DNA, and then we must look at the methods used to identify overlapping clones and hence build up the clone contig.

4.2.4 Cloning Vectors for Large Pieces of DNA

The first cloning vectors to be developed, back in the 1970s, were based on bacterial plasmids. These types of plasmid vector are still used as workhorses in general DNA manipulation but they are unstable if their total size is increased by more than a few kb, so they cannot be used to clone large fragments of DNA and are unsuitable for the purpose that we are discussing. The desire to work with larger pieces of DNA led to the development of cloning vectors based on bacteriophage λ, including both regular λ vectors and the λ-plasmid hybrids called *cosmids*. With these vectors, the upper limit for the length of the cloned DNA is set by the space available within the head of the λ phage particle, as the recombinant vector DNA must be packaged within λ phages in order to be transferred to new *E. coli* cells during the cloning process. There is only so much space in the λ phage head so these vectors cannot be improved in any way: their maximum capacities have already been reached. Despite this limitation, some small eukaryotic genomes, notably that of the yeast *Saccharomyces cerevisiae,* have been sequenced mainly or entirely from clone contigs constructed with cosmids.

The first major breakthrough in attempts to clone DNA fragments larger than 100 kb came with the invention of *yeast artificial chromosomes* or *YACs*. These vectors are propagated not in *E. coli* but in *S. cerevisiae* and they are based not on plasmids or viruses but on chromosomes. The first YACs were constructed after studies of natural chromosomes had shown that, in addition to the genes that it carries, each chromosome has three important components:

1. The *centromere*, which plays a critical role during nuclear division.
2. The *telomeres,* the special sequences which mark the ends of chromosomal DNA molecules.
3. One or more *origins of replication*, which initiate synthesis of new DNA when the chromosome divides.

In a YAC, the DNA sequences that underlie these chromosomal components are linked together with one or more selectable markers and at least one restriction site into which new DNA can be inserted.

All of these components can be contained in a DNA molecule of 10-15 kb in length. As natural yeast chromosomes range from 230kb to over 1700kb, the possibility arises that a YAC could be used to clone Mb-sized DNA fragments. This potential has been realized, standard YACs being able to clone 600 kb fragments, with special types able to handle DNA up to 1400 kb in length. Currently this is the highest capacity of any type of cloning vector and several genome projects have made extensive use of YACs. Unfortunately, with some types of YAC there have been problems with insert stability, the cloned DNA sometimes becoming rearranged by recombination. For this reason there is also great interest in other types of vectors, ones that cannot clone such large pieces of DNA but which do not suffer so much from instability problems. These vectors include:

1. *Bacteriophage P1* vectors, which are very similar to λ vectors, being based on a deleted version of a natural phage genome, the capacity of the cloning vector being determined by the size of the deletion and the space within the phage particle. The P1 genome is larger than the λ genome, and the phage particle is bigger, so a P1 vector can clone larger fragments of DNA than λ, up to 125 kb using current technology.
2. *Bacterial artificial chromosomes or BACs* are based on the naturally-occurring F plasmid of *E. coli.* Unlike the plasmids used to construct the early cloning vectors, the F plasmid is relatively large and vectors based on it therefore have a higher capacity for accepting inserted DNA. BACs can be used to clone fragments of 300 kb and greater.
3. *PI-derived artificial chromosomes* or *PACs* combine features of P1 vectors and BACs and have a capacity of up to 300 kb.
4. *Fosmids* contain the F plasmid origin of replication and a λ *cos* site. They are similar to cosmids but have a lower copy number in *E. coli,* which means that they are less prone to instability problems.

4.2.4.1 Clone contigs can be built up by chromosome walking, but the method is laborious

The simplest way to build up an overlapping series of cloned DNA fragments is to begin with one clone from a library, identify a second clone whose insert overlaps with the insert in the first clone, then identify a third clone whose insert overlaps with the second clone, and so on. This is the basis of *chromosome walking,* which was the first method devised for assembly of clone contigs.

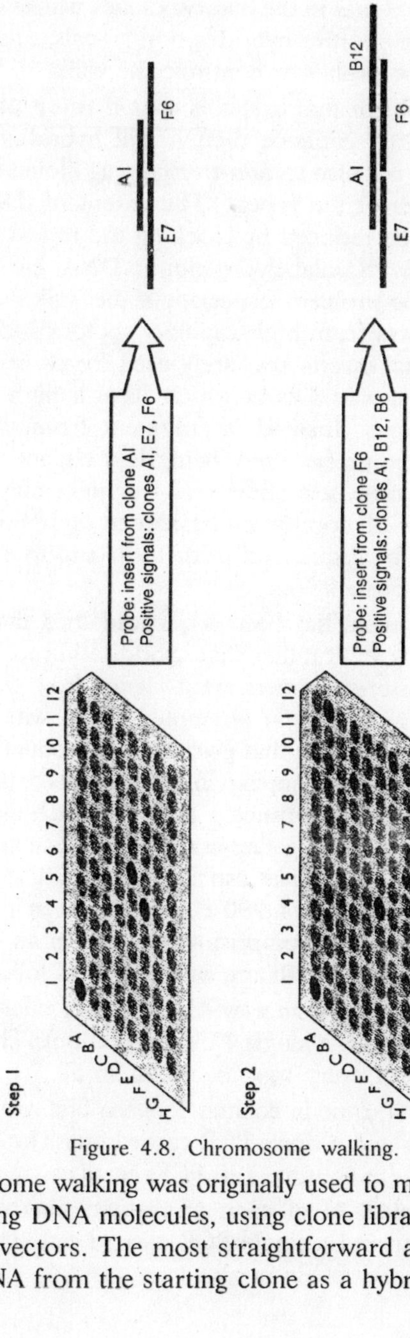

Figure 4.8. Chromosome walking.

Chromosome walking was originally used to move relatively short distances along DNA molecules, using clone libraries prepared with λ or cosmid vectors. The most straightforward approach is to use the insert DNA from the starting clone as a hybridization probe to screen all the other clones in the library. Clones whose inserts overlap with the probe give positive hybridization signals, and their inserts can be used as new probes to continue the walk.

The main problem that arises is that if the probe contains a genome-wide repeat sequence then it will hybridize not only to overlapping clones but also to non-overlapping clones whose inserts also contain copies of the repeat. The extent of this nonspecific hybridization can be reduced by blocking the repeat sequences by prehybridi-zation with unlabeled genomic DNA, but this does not completely solve the problem, especially if the walk is being carried out with long inserts from high-capacity vectors such as YACs or BACs. In fact, intact inserts are rarely used for chromosome walks with human and similar DNAs which have a high frequency of genome-wide repeats. Instead, a fragment from the end of an insert is used as the probe, there being less chance of a genome-wide repeat occurring in a short end-fragment compared to the insert as a whole. If complete confidence is required then the end-fragment can be sequenced prior to its use to ensure that no repetitive DNA is present.

If the end-fragment has been sequenced then the walk can be speeded up by using PCR rather that hybridization to identify clones with overlapping inserts. Primers are designed from the sequence of the end-fragment and used in attempted PCRs with all the other clones in the library. A clone that gives a PCR product of the correct size must contain an overlapping insert. To speed the process up even more, rather than performing a PCR with each individual clone, groups of clones are mixed together in such a way that an unambiguous identification of overlapping ones can still be made. The method stated already in which a library of 960 clones has been prepared in 10 microtiter trays, each tray comprising 96 wells in an 8 × 12 array, with one clone per well. PCRs are carried out as follows:

1. Samples of each clone in row A of the first microtiter tray are mixed together and a single PCR carried out. This is repeated for every row of every tray, 80 PCRs in all.
2. Samples of each clone in column 1 of the first microtiter tray are mixed together and a single PCR carried out. This is repeated for every column of every tray, 120 PCRs in all.
3. Clones from well Al of each of the ten microtiter trays are mixed together and a single PCR carried out. This is repeated for every well, 96 PCRs in all.

As explained in the legend these 296 PCRs provide enough information to identify which of the 960 clones give products and which do not. Ambiguities arise only if a substantial number of clones turn out to be positive.

4.2.5 Newer, More Rapid Methods for Clone Contig Assembly

Even when the screening step is carried out by the combinatorial PCR approach chromosome walking is a slow process and it is rarely possible to assemble. contigs of more than 15-20 clones by this method. The procedure has been extremely valuable in *positional cloning*, where the objective is to walk from a mapped site to an interesting gene that is known to be no more than a few Mb distant. It has been less valuable for assembling clone contigs across entire genomes, especially with the complex genomes of higher eukaryotes. So what alternative methods are there?

The main alternative is to use a *clone fingerprinting* technique. Clone fingerprinting provides limited information on the physical structure of a cloned DNA fragment, this physical information or 'fingerprint' being compared with equivalent data from other clones, enabling those with similarities, possibly indicating overlaps, to be identified. One or a combination of the following techniques are used:

1. *Restriction patterns* can be generated by digesting clones with a variety of restriction enzymes and separating the products in an agarose gel. If two clones contain overlapping inserts then their restriction fingerprints will have bands in common, as both will contain fragments derived from the overlap region.
2. *Repetitive DNA fingerprints* can be prepared by blotting a set of restriction fragments and carrying out Southern hybridization with probes specific for one or more types of genome-wide repeat. As for the restriction fingerprints, overlaps are identified by looking for two clones that have some hybridizing bands in common.
3. *Repetitive DNA PCR, or interspersed repeat element PCR* (IRE-PCR), uses primers that anneal within genome-wide repeats and so amplify the single copy DNA between two neighbouring repeats. Because genome-wide repeat sequences are not evenly spaced out in a genome, the sizes of the products obtained after repetitive DNA PCR can be used as a fingerprint in

comparisons with other clones, in order to identify potential overlaps. With human DNA, the genome-wide repeats called *Alu* elements are often used as these occur on average once every 4 kb. An *Alu-PCR* of a human BAC insert of 150 kb would therefore be expected to give approximately 38 PCR products, producing a detailed fingerprint.

4. STS content mapping is particularly useful because it can result in a clone contig that is anchored on to a physical map of STS locations. PCRs directed at individuals STSs are carried out with each member of a clone library. Presuming the STS is single copy in the genome, then all clones that give PCR products must contain overlapping inserts.

As with chromosome walking, efficient application of these fingerprinting techniques requires combinatorial screening of gridded clones, ideally with computerized methodology for analyzing the resulting data.

4.2.6 Directed Shotgun Approach

Experience with the conventional shotgun approach to genome sequencing has shown that if the total length of sequence that is generated is between 6.5 and 8 times the length of the genome being studied, then the resulting sequence contigs will span over 99.8% of the genome sequence, with a few gaps that can be closed by one of the methods developed during the H. *influenzae* project. This implies that 70 million individual sequences, each 500bp or so in length, corresponding to a total of 35000 Mb, would be sufficient if the random approach were taken with the human genome. Seventy million sequences is not an impossibility: in fact, with 75 automatic sequencers, each performing 1000 sequences per day, the task could be achieved in 3 years.

The big question is whether the 70 million sequences could be assembled correctly. If the conventional shotgun approach is used, which makes no reference to a genome map, then the answer is certainly no. The huge amount of computer time needed to identify overlaps between the sequences, and the errors or at best uncertainties caused by the extensive repetitive DNA content of the human genome, would make the task impossible. But it has been argued that a 'directed shotgun' approach, one which makes full use of the genome map during the sequence assembly phase of the project, would work.

This project is currently underway with the 70 million sequences being generated as follows:

1. Sixty million sequences are being obtained from a plasmid library of human DNA with inserts averaging 2 kb in length. Approximately 500 bp are being sequenced from both ends of the inserts in these clones.
2. Ten million sequences are being obtained as end sequences from a second library, prepared with a different type of plasmid cloning vector, containing 10-kb inserts.

Two libraries are being used because with any cloning vector it is anticipated that some fragments will not be cloned because of incompatibility problems which prevent vectors containing these fragments from being propagated. Different types of vector suffer

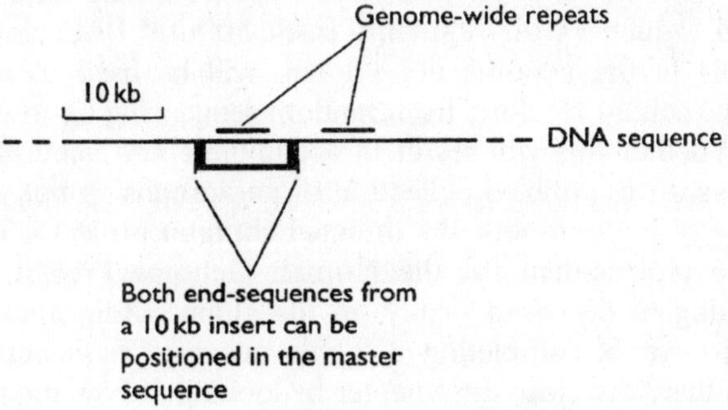

(B) Incorrect sequence assembly

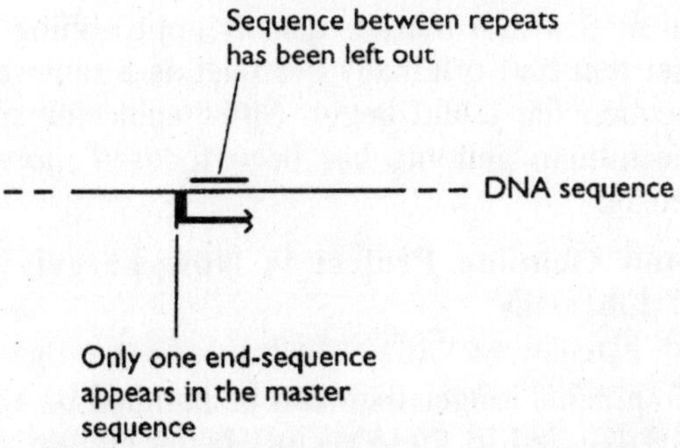

Figure 4.9. Avoiding errors when the directed shotgun approach is used.

from different problems so fragments that cannot be cloned in one vector can often be cloned if a second vector is used. Using two different vectors should therefore improve the overall coverage of the genome. The 10-kb library will also help to ensure that errors do not occur when sequencing around the commonest genome-wide repeats. Most of these repeats are 5 kb or so in size, and hence can be entirely contained in a 10-kb fragment but not a 2-kb one. Sequence 'jumps', from one repeat sequence to another, can therefore be avoided by ensuring that the two endsequences of each 10-kb insert are at their appropriate positions in the master sequence.

The critical feature of this project is that the master sequence will be assembled with reference to the framework provided by the human genome maps. In particular, the STS map data as well as the end sequences of fragments from 300000 BAC clones whose positions in the genome are known, will be used to anchor the sequence contigs resulting from random sequencing on to the genome map. Whether this will result in a complete and accurate genome sequence, or a 'jumbled collection of paragraphs' is not yet known. Because of these doubts, the directed shotgun project is not looked on as a replacement for the Human Genome Project, the latter continuing to be based firmly on the clone contig approach with the objective of completing a highly accurate sequence by 2003. We will therefore close this chapter by looking at how the sequencing phase of the Project is progressing.

4.3 SEQUENCING PHASE OF HUMAN GENOME PROJECT

We surveyed the mapping phase of the Human Genome Project and saw how this culminated in 1995 with an integrated genetic and physical map with a marker density approaching one per 100 kb, the target that had originally been set as a requirement before largescale sequencing could begin. Since achieving this objective, work on the human genome has been focused more directly on DNA sequencing.

4.3.1 Human Genome Project is Now Largely Based on BAC Libraries

The high capacity of YACs, which means that they can be used with DNA fragments longer than can be handled by any other type of cloning system, led to these vectors being chosen as the source material for construction of the first clone contigs of the entire human genome and of individual chromosomes. The original idea was that the YAC libraries would also be used for the sequencing phase of the Project, but this plan has largely been abandoned because of concerns that some YAC clones (up to half according to some estimates) contain noncontiguous fragments from different parts of the genome or have undergone rearrangements or deletions. These instabilities would lead to major errors if YAC inserts were used to generate the genome sequence.

To avoid these errors, a library of 300000 BAC clones has been generated and these clones mapped onto the genome. This forms the 'sequence-ready' map which is acting as the primary foundation for the sequencing phase of the project, the inserts from each BAC being completely sequenced by the shotgun method.

4.3.2 Rapid Sequence Acquisition can be Directed at Interesting Regions of the Genome

It is recognized that the clone contig approach being taken by the Human Genome Project will result in a relatively slow accumulation of sequence data. With the complete genome sequence not due until 2003, a number of alternative strategies, designed to generate sequence data more quickly than is possible by rigorous mapping and contig assembly, have been developed. Bear in mind that although the genome as a whole is interesting, the very interesting bits are the genes, especially those that are relevant to human disease. Rapid sequencing strategies that result specifically in gene sequences are therefore worth pursuing.

4.3.3 EST Sequencing is a Rapid Way of Accessing Genes

Genes can be directly accessed by analysis of expressed sequence tags (ESTs). These are a type of STS that can be used as markers in physical mapping. ESTs are obtained by sequencing cDNA, which itself is synthesized from the mRNA molecules in a cell. The mRNAs in a cell are copies of the genes that are being expressed, and do not contain sequences from the regions between genes, nor from the noncoding introns that are present within many genes. ESTs are therefore copies of just the very interesting parts of the genome. ESTs also have a number of practical advantages which mean that their sequences can be generated very rapidly:

1. mRNA is converted into cDNA by a simple test-tube reaction, so a cDNA library can be prepared very quickly.

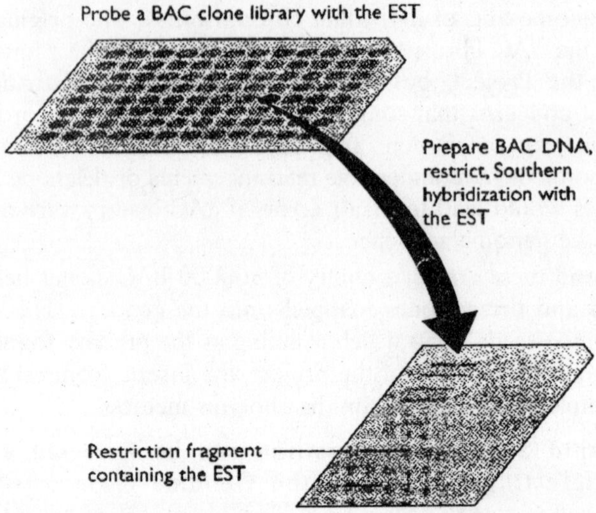

Figure 4.10. From EST to gene sequence.

2. Only one sequencing experiment is needed per cDNA, because 500 bp of sequence is enough to identify the gene that an EST represents, though in practice both end-sequences are usually obtained.
3. The sequence of an EST does not need to be checked by repeating the sequencing experiment, because mistakes arising from the inherent error rate of chain termination sequencing (about one error per 100 nucleotides) do not prevent identification of the gene from which the EST is derived. In contrast, the minimal acceptable error rate for the genome sequence (one per 10000 nucleotides, as set by the Human' Genome Project) requires that each region be sequenced at least three times.

By late 1998, over one million human ESTs had been obtained, representing 50000 to 60000 different human genes. In other words, partial sequences are available for well over half of all human genes. The fact that these are only partial sequences, either because the sequencing experiment did not cover the entire cDNA, or the cDNA synthesis did not result in a complete copy of the mRNA, is not a problem. As mentioned above, the sequence information in a single EST is enough to identify the gene from which it is derived, nonidentifications resulting not from a lack of sequence data but from limitations in the procedures used for gene identification. If the complete gene sequence is needed then the EST can be used as a hybridization probe, first to find the appropriate clone in a BAC library, and then to identify the region within the clone that contains the gene sequence.

4.3.4 Sequence Skimming is Equally Rapid

A second method of rapid sequence acquisition is *sequence skimming*, which in effect is the first stage of a shotgun sequencing project. The rationale is that the first few random sequences obtained from a long fragment will map at various positions within that fragment, so if the fragment contains one or more genes then there is a good chance that at least some of them will be 'hit' by these initial sequences. For example, imagine the fragment being studied is 100 kb in length. One hundred random sequences of 500bp each will produce about 50 kb of sequence, which even allowing for some overlaps would be anticipated to include over one-third of the 100 kb fragment. These initial sequences will be error prone, but as with ESTs this is not a problem if the initial objective is simply to identify the genes that are present.

4.3.5 Sequencing Phase of the Human Genome Project Raises Ethical Questions

The Human Genome Project has been controversial since its inception and now that large-scale sequencing is getting underway a number of ethical issues are being raised. One controversy concerns the question of who if anyone will own human DNA sequences. To many, the idea of ownership of a DNA sequence is a peculiar concept, but large sums of money can be made from the information contained in the human genome, for example by using gene sequences to direct development of new drugs and therapies against cancer and other diseases. Pharmaceutical companies involved in genome sequencing naturally want to protect their investments, as they would for any other research enterprise, and currently the only way of doing this is by patenting the DNA sequences that they discover. Unfortunately, in the past, errors have been made in dealing with the financial issues relating to research into human biological material, the individual from whom the material is obtained not always being a party in the profit-sharing. These issues have still to be resolved.

The problems relating to the public usage of human genome sequences are even more contentious. A major concern is the possibility that once the sequence is known, individuals whose sequences are considered 'substandard', for whatever reason, might be discriminated against. The dangers range from increased insurance premiums for individuals whose sequence includes mutations predisposing them to one or other genetic disease, to the possibility that racists might attempt to define good and bad sequence features, with depressingly predictable implications for the individuals unlucky enough to fall into the 'bad' category. Many scientists argue in support of a Human Genome Diversity Project, which will build up a catalog of sequence variability in different populations and ethnic groups, in the expectation that the resulting data will emphasize the unity of the human race by showing that patterns of genetic variability do not reflect the geographical and political groupings that humans have adopted during the last few centuries.

The Human Genome Project, especially in the USA, continues to support research and debate into the ethical, legal and social issues raised by genome sequencing. In parti-cular, great care is being taken to ensure that the genome sequence that results from the project cannot be identified with any single individual. The DNA that is being cloned and sequenced is taken only from individuals who have given full consent for their material to be used in this way and for whom anonymity can be guaranteed. This has required a certain amount of realignment of thc research effort, as older clone libraries have been destroyed and the existing physical maps have had to be checked using the new material. It is accepted, however, that this extra work is necessary and that care must be taken to maintain and enhance public confidence in the project.

5

Understanding Proteomics

Function prediction from sequence employs a combination of highly efficient sequencing techniques and sophisticated electronic searches against millions of gene and protein sequences. By providing access to biological databases and powerful tools for the prediction of the biochemical functions and cellular roles of gene products, bioinformatics can inform and focus research and enhance its efficiency. Therefore, experimental and computational methods are becoming closely integrated in biological and biomedical investigations. This integration becomes possible when bioinformatics techniques are not only used at *isolated* points in *the* discovery process, but when *theoretical* predictions are *directly* tested in the laboratory and results are fed back for creating refinements in theoretical models. The refined model in turn becomes the starting point for further experiments, and new results then lead to further model refinement. Such an iterative approach not only supports efficiency, but also enhances the understanding of the biological system under study.

5.1 SIMILAR SEQUENCE-SIMILAR STRUCTURE-SIMILAR FUNCTION PARADIGM

Bioinformatics seeks to elucidate the relationships between biological sequence, threedimensional structure and its accompanying functions, and then to use this knowledge for predictive purposes. The guiding principles for the study of sequence-structure-function relationships, and the predictive techniques resulting from this analysis, are derived from the processes of molecular evolution. Most commonly, protein function is inferred from the known functions of homologous proteins; i.e., proteins predicted to share a common ancestor with the query protein based on significant sequence similarity. For homologous proteins with easily recognizable sequence similarity, this type of prediction is based on the *'similar sequence-similar structure-similar function'* *paradigm*.

The rule assumes a one-to-one relationship between protein sequence and unique 3-D structure, and between structure and unique biochemical function. Whilst this usually holds true for closely related proteins, it is important to be aware that more distant relatives belonging to protein families typically exhibit a complex pattern of shared and distinct characteristics in terms of biochemical function, ligand specificity, gene regulation, protein-protein interactions, tissue specificity, cellular location, developmental phase of activity, biological role, and other features.

One can expect to find an entire spectrum of function conservation in evolutionarily related genes, ranging from *paralogs* with a very high degree of functional divergence to others with strictly conserved function, the latter most likely being *orthologs* involved in core biological processes. *Sequence alignments* provide a powerful technique to compare novel sequences with previously characterized proteins. If a homologous relationship between query and target protein can be inferred based on significant sequence similarity over most of the aligned sequence, it is often possible to make quite detailed predictions regarding biochemical function and cellular location, but the biological processes the respective proteins are involved in may not be identical. In contrast, if only a significant local match to one or more protein domains is identified, function assignment for the entire protein will be less reliable. Ideally, function prediction by inferred homology ought to be confirmed by appropriate experimental analysis.

5.2 FUNCTIONAL ANNOTATION OF BIOLOGICAL SEQUENCES

5.2.1 Identification of Sequence Homologs

Functional and evolutionary information can be inferred from *sequence comparisons*. These methods are most sensitive at the protein level and the detection of distantly related sequences is easier in protein translation. The reason for this is the redundancy of the genetic code. Nucleotide sequences therefore first need to be automatically translated into their predicted primary amino acid sequence. The first step toward function identification generally involves searching protein sequence databases with a query sequence using automatic pairwise comparison tools such as *BLAST* or *FASTA*.

Position specific iterative BLAST (*PSI-BLAST*), developed by Altschul and co-workers, can be employed to extend the search to distantly related homologs. The PSI-BLAST server at the NCBI website is easy to use and for most protein families returns a list of putative relatives within minutes of requesting the search. The PSI-BLAST algorithm is also relatively easy to implement locally in order to run against a user's database. PSI-BLAST accepts a query sequence as input and constructs a *profile* (or *position-specific scoring matrix, PSSM*) from a multiple alignment of the highest scoring hits in an initial gapped BLAST search.

The algorithm calculates position-specific scores for every alignment position; highly conserved positions are assigned high scores and weakly conserved positions receive scores close to zero. Highly conserved positions would be expected to be present in all members of the family, and information about residue preferences at these conserved positions are therefore diagnostic for the family and can be encoded in a family profile. PSI-BLAST uses a substitution matrix which assigns a score for aligning any possible pair of residues based on the frequency with which that substitution is known to occur among homologous proteins. Different substitution matrices are designed to detect similarities at certain ranges of evolutionary distance.

Experimental validation showed that the BLOSUM62 matrix is very well suited to detect distant evolutionary relationships. This substitution matrix is therefore implemented as the default in PSI-BLAST, although other matrix options are also included. The profile is used to identify more distant relatives in an iterative process which results in increased sensitivity. A conservative threshold is applied to minimize the probability of including any unrelated sequences of false positives. In each iteration, the profile is refined by including the new members until the process converges on a unique set of proteins and no new sequences are found. For some families this may involve very many cycles, with the danger that the profile drifts beyond the actual variability for the particular family, by including false positives. Therefore, it is possible to stop the algorithm after a fixed number of cycles, typically 20, to minimize the risk of profile drift. When performing searches with BLAST, FASTA or PSI-BLAST, a statistically significant match to a well-characterized protein over the entire length of the query sequence may be returned. However, in many cases, searches will only identify local similarities between the query sequence and a number of diverse proteins that may be otherwise unrelated. These local matches often signal the presence of shared homologous domains.

5.2.2 Identification of Conserved Domains and Functional Sites

Larger proteins are modular in nature, and their structural units, protein domains, can be covalently linked to generate multi-domain proteins. *Domains* are not only structurally, but also functionally, discrete units, i.e., specific biochemical functions are carried out by specific domains. Domain family members are structurally and functionally conserved and recombined in complex ways during evolution. Novelty in protein function often arises as a result of the gain or loss of domains, or by re-shuffling existing domains along the linear amino acid sequence. Domains can be seen as the 'units of evolution', and, therefore, both structural and functional similarity between proteins needs to be analyzed at the domain level. For pairs of protein domains that share the same three-dimensional fold, precise function appears to be conserved down to ~40% sequence identity, whereas broad functional class is conserved to ~25% identity.

When conducting sequence database searches with BLAST or FASTA, local matches to domains present in multi-domain proteins need to be evaluated with caution. Firstly, the domains that correspond to the query need to be correctly identified and, secondly, it is important to be aware that domain function can vary significantly depending on context. Automatic transfer of functional annotation can be misleading. The relationship between members of domain families is analogous to the relationship between genes belonging to protein families: Database search algorithms cannot distinguish between a match to an *ortholog* or a *paralog.* Since this distinction cannot be derived automatically with the available algorithms, the returned matches need to be carefully evaluated.

The higher the level of sequence conservation between genomes, the more likely it is that the query and the matched sequence are orthologs (operationally defined as symmetric top-scoring protein sequences in a sequence homology search). Current *global* and *local sequence alignment* methods, and consensus methods, are reliable for identifying sequence relatedness when the level of sequence identity is above 30%.

Consensus methods occupy an intermediate place between multiple alignment and pattern/profile methods. Consensus sequences are built from multiple alignments of related sequences and the diagnostic sequence motif is represented as the majority or plurality character at each position. This short sequence alignment illustrates this method:

DLLFRCG
ELLAKCE
ELVFRCG
DILFKCA
DLLFRCG

DLLFRCG = consensus

Query sequences are then matched against a database of consensus sequences by BLAST. The *Protein Domain Database (ProDom),* a secondary database, was constructed by clustering homologous sequence segments derived from SWISS-PROT and provides multiple alignments and consensus sequences for homologous domain families. Whilst consensus sequences make some attempt at representing the pattern of mutational change at each position of the alignment to increase sensitivity, information about rarely occurring amino acids is lost and the sensitivity of such methods does not extend below 30% sequence identity. In order to detect signals of more remote relationships in protein sequences, below a level of ~30% sequence identity, different analytical approaches have been developed which are based on representations of similarity at the level of functional protein domains. Below 20% sequence identity, evolutionary relationships are difficult to elucidate without a combination of structural and functional evidence to infer homology. Therefore, purely sequence-based comparisons fail to identify many of the evolutionary relationships that emerge once the structures of proteins are compared.

The BLAST algorithm finds only 10% of distant relationships in the Protein Data Bank (PDB) of three-dimensional structures, and although much more sensitive, PSI-BLAST also misses many of these relationships. To allow the *identification of homologous domains* with low sequence identity (<30%), *pattern and profile methods* enhance sequence information by a mutation pattern at a given position along the sequence that is derived from an alignment of homologous proteins. For the purpose of function assignment,

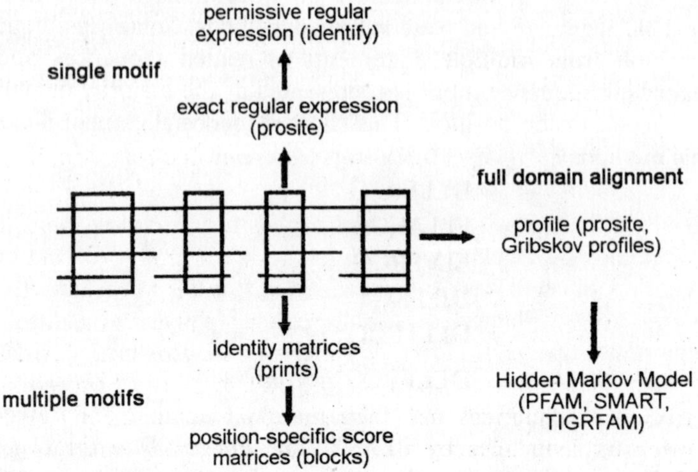

Figure 5.1. Methods for building macromolecular motif databases. The direction of the arrows indicates increased sensitivity and decreased specificity of the representation.

the concept of a domain is defined broadly as a region of sequence homology among sets of proteins that are not all full-length homologs.

Homology domains often, but not always, correspond to recognizable protein-folding domains. In multiple alignments of distantly related sequences, highly conserved regions, called *motifs,* features, signatures, or blocks, are discernible, surrounded by divergent regions exhibiting a high degree of mutational change among family members. Motifs are typically ~10-20 amino acids in length and tend to correspond to core structural and functional elements of the proteins. *Single and multiple motif methods* (regular expressions, identity and position-specific score matrices) utilize motif regions in multiple alignments to build patterns diagnostic of a particular protein family.

Full domain alignment methods (profiles and Hidden Markov Models) create complete representations of all features in the aligned sequences of a domain family (all highly conserved and divergent regions in a complete domain alignment) and thereby maximize the overall signal of the domain family. Using pattern and profile representations, databases of diagnostic protein family motifs have been created and can be searched with a query sequence. This may reveal the presence of one or several regions corresponding to particular functions in the query sequence. Pattern and profile database searches therefore constitute an important resource for function prediction. These databases are also known as *secondary databases,* because their models of biological motifs were generated using the data contained in primary sequence databases.

5.2.3 Methods for Building Macromolecular Motif Representations

There are two general approaches to improving the sensitivity of homology searching by using family information. One of these restricts motif models to key conserved features to exclude the higher noise level of the more variable regions (single and multiple motif methods), the other approach includes all possible information to maximize the overall signal of an entire protein or protein domain (full alignment methods). Both approaches are valid. Within these two groupings, numerous representation methods exist and are used alone or in combination by different secondary databases. No fixed rule can be given for choosing the method to use in the analysis of a given protein sequence, as each protein family has a different conservation pattern.

In practice one may decide which method is most appropriate if the protein family is known, but it is advisable to perform both types of searches. It is also important to keep in mind that linear sequences constitute highly abstract models. Whilst the amino acid composition of proteins, their domains and functional sites, can be represented in this fashion, biochemical functionality of motif regions can only be fully understood when taking into account the three-dimensional arrangement of the involved amino acid residues within the context of a protein's structure. It is therefore instructive, and essential, to relate one-dimensional sequence models to protein structures if available. In so doing, we gain a clearer picture of what is encoded in motif patterns.

Figure 5.2 shows an example of two carboxypeptidase structures; one is a monomeric prohormone-processing carboxypeptidase from the yeast *Saccharomyces cerevisiae,* kexl(delta)p (PDB code 1ac5), the other is a dimeric human cathepsin a (POB code 1ivy). Both proteins belong to the family of serine histidine carboxypeptidases, which can be identified from their sequences by the presence of a diagnostic pattern that maps to the active-site. The PROSITE secondary database contains an entry (PS00560) that encodes the motif in a pattern called a regular expression: [LIVF]-x(2)-[LIVSTA] -x -[IVPST]-x -[GSONQL]-[SAGV]-[SG] -H-x-[IVAQ]-P-x(3)-

[LIVF]-x(2)-[LIVSTA]-x-[IVPST]-x-[GSDNQL]-[SAGV]-[SG]-H-x-[IVAQ]-P-x(3)-[PSA]

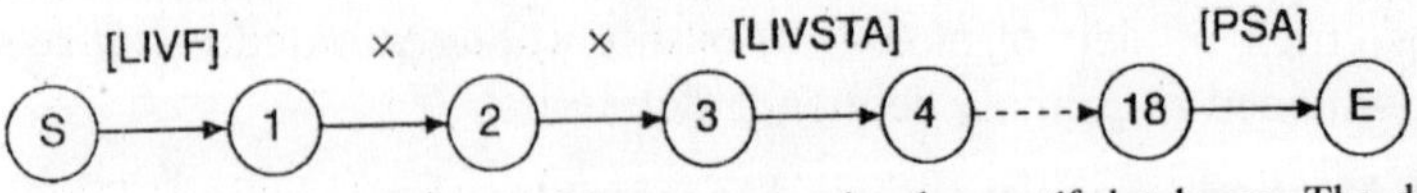

Figure 5.2. Methods for building macromolecular motif databases. The direction of the arrows indicates increased sensitivity and decreased specificity of the representation.

[PSA]. The sequence regions that match this pattern in each of the proteins are shown below the structural models of the active-sites with the motif residues colour-coded as in the sequences. As will be explained below, the regular expression matches each of the regions exactly, whilst both sequences considerably differ from each other.

5.2.3.1 Single motif methods

5.2.3.1.1 Regular expressions

Sequence patterns representing a single motif can be encoded in *regular expressions* which represent features by logical combinations of characters. The secondary database *PROSITE* encodes some motifs as regular expressions. The basic rules for regular expressions are:

- Each position is separated by a hyphen '-'
- An uppercase character matches itself
- x means 'any residue'
- [ ] surround ambiguities—a string [LIV] matches any of the enclosed symbols
- A string [R]* matches any number of strings that match
- { } surround forbidden residues
- ( ) surround repeat counts.

Patterns encoded in regular expressions are restricted to key conserved features in order to reduce the 'noise' of more variable regions. Regular expressions are built by hand in a step. wise fashion from multiple alignments of protein families. Typically, the first step is to encode functionally significant residues, such as the catalytic site or a post-translational modification site, in a core pattern, usually 4-5 residues long. Subsequently, the length of the pattern is gradually increased to about 10-20 residues. Using regular expressions, it is possible to include gaps in a pattern.

At times, a highly conserved region without known functional significance can also be diagnostic for a given protein family and may be encoded in a pattern. Such regions may be essential in maintaining the stability of the shared threedimensional fold of protein families and thus constitute structural signature motifs. Regular expressions, such as PROSITE patterns, are matched to primary amino acid sequences using 'finite state automata'. Thc UNIX pattern matching (word search) utilities such as *grep, sed, awk* and *vi* are examples of highly efficient implementations of finite state automata.

A finite state automaton is a model composed of a number of states, and the states are interconnected by state transitions. It reads one symbol at a time from an input string. The symbol may be accepted, the automaton then enters a new state; or the symbol may not be accepted in which case the automaton stops and rejects the string. *Scoring* against a regular expression is therefore *binary* (YES/NO), the pattern either matches the sequence or it does not. If the automaton reaches a final accepting state, the input string has been successfully recognized. Fig. 5.2. depicts a finite state automaton for matching the PROSITE CARBOXYPEPT _SER_HIS (PS00560) regular expression.

To check whether a sequence matches this expression, the sequence is fed to the automaton one symbol at a time (S denotes the start site). If the first symbol matches the expression [LIVF], the automaton enters state 1; otherwise it quits and rejects the sequence. If the automaton is in state 1 and it reads a symbol that matches x (any valid amino acid symbol), it moves to state 2; when in state 2, it again reads any valid amino acid symbol, it moves to state 3; when in state 3, it reads a symbol that matches the expression [LIVSTA], it moves to state 4, and so on until it successfully matches the sequence to the whole regular expression by reaching the end state E.

Regular expressions perform best when a given protein family can be characterized by a highly conserved single motif. They can also be used to encode short sequence features believed to be the result of convergent evolution, such as sugar attachment sites, phosphorylation sites, acetylation sites, etc. However, the caveat applies that the shorter the pattern (maybe as short as only 3-5 residues) the more likely purely random matches become. Such matches are inherently unreliable and can only provide a hint that such a site may be present. Regular expressions possess a number of limitations. Firstly, they lose information, as only those features that are judged to be the most conserved or significant from a parent alignment are modeled. However, as more family members are identified, alignments will become more and more representative of the family and motifs earlier thought to be characteristic of the family might have to be changed considerably or completely replaced by more significant ones.

Secondly, regular expressions are built from an incomplete sample of all possible members of a protein family and only observed amino acid variability is encoded. The substitution probabilities of unobserved amino acids are not modeled.

Thirdly, matches have to be exact. Even one mismatch will cause the query to be rejected, although the mismatch may be a conservative amino acid substitution or an insertion/deletion with conserved function. Regular expressions will therefore fail to detect new family members that match a partial motif (false-negative hits). Lastly, unlikely sequences may be accepted if patterns are made too inclusive in an attempt to detect distant relatives (false-positive hits).

5.2.3.1.2 Fuzzy regular expressions

More permissive regular expressions can be built by including information on shared biochemical properties of amino acids. Amino acids can be grouped according to hydrophobicity, charge, size, presence of aromatic rings etc., and these groupings can then be used to derive *fuzzy regular expressions* aimed at the detection of more distant relatives. The inherent disadvantage of this method is the increased likelihood of chance matches without any biological significance. An example illustrates this approach:

Alignment	A possible fuzzy regular expression
FKA...	
YPI...	[FYW] – [PKR] – [VIALG]...
FPV...	
FKV...	

Fuzzy regular expressions can be built based on different amino acid groupings, and, if constructed by hand, also contain an element of subjective judgement as to how permissive one wants the pattern to be. The same alignment can be the starting point for quite different fuzzy regular expressions. For the first column of the example alignment above, the third aromatic amino acid tryptophan (W), although unobserved, was included in the regular expression, because both phenylalanine (F) and tyrosine (Y) are aromatic residues; for the second column, arginine (R) was included because it shares a positive charge with lysine *(K);* for the third column, leucine (L) was included as it completes the set of aliphatic hydrophobic residue together with valine (V) and isoleucine (I), and glycine (G) shares the property of smallness with alanine (A). Fuzzy regular expressions, built from source alignments held in the *BLOCKS* and *PRINTS* databases, are employed by the *eMOTIF/ IDENTIFY* sequence annotation system.

5.2.3.2 Multiple motif methods

In most protein family alignments, more than one conserved motif region will be present. *Multiple motif methods* represent several or all of these family-specific regions in order to model the mutual context provided by motif neighbours and thereby improve the diagnostic power of the pattern. The query sequence of a more distant relative may only match a subset of motifs, but the pattern of matched motifs may still allow statistically significant, and biologically meaningful, family assignments to be made.

5.2.3.2.1 Fingerprints

The *PRINTS* database encodes multiple motifs (called *fingerprints*) in ungapped, unweighted local alignments which, unlike regular expression or consensus methods, preserve all sequence information in aligned blocks. Usually the motifs do not overlap, but are separated along a sequence, though they may be contiguous in the three-dimensional protein structure. An advantage of providing actual alignments is that they contain all of the information needed to derive other representations. A variety of different scoring methods may then be used, such as BLAST searching, providing different perspectives on the same underlying data.

Fingerprints are built manually by initially starting from a small multiple alignment which provides the seed motifs. Subsequent iterative database scanning will create an increasingly informative, composite set of motifs by retrieving, at each round, all additional sequences that match all the motifs and can be used to improve the fingerprint. This iterative process is repeated until no further complete fingerprint matches can be identified.

5.2.3.2.2 BLOCKS

The *BLOCKS* database is derived from *PROSITE* and *PRINTS* entries. Using the most highly conserved regions in protein families contained in PROSITE, a motif-finding algorithm first generates a large number of candidate *blocks.* Initially, three conserved amino acid positions anywhere in the alignment are identified and such a spaced triplet, if found in enough sequences, anchors the local multiple alignment against which sequences lacking the triplet are aligned to maximize a block's sensitivity. In this fashion, blocks are iteratively extended and ultimately encoded as ungapped local alignments. Graph theory techniques are used to assemble a best set of blocks for a given family. After a block is made, weights are computed for each sequence segment contained in it. These weights compensate for overrepresentation of some of the sequences; low weights are given to redundant sequences, and high weights to diverged sequences.

The BLOCKS database uses position-based sequence weights, which are easy to compute and have been shown to perform well in searches. These are simple weights derived from the number of different residues and their frequency in each position in the block. The BLOCKS search system, allowing query sequence searches against the database, converts each block to a *position-specific scoring matrix* (*PSSM*), which is similar to a profile. A PSSM has as many columns as there are positions in the block, and 20 rows, one for each amino acid. It also contains additional rows for special characters: B stands for the amino acids aspartic acid (D) or asparagine (N); Z for glutamic acid (E) or glutamine (Q); X for an unknown amino acid residue; and—denotes a gap.

Each PSSM entry consists of numeric scores that are based on the ratio of the observed frequency of an amino acid in a block column to its expected overall frequency in SWISS-PROT (*odds ratio*). The observed frequencies are weighted for sequence redundancy using the sequence weights in the block. However, it is typical that most amino acids do not occur at all in a position, and so one must decide *on* a method to deal with these unobserved residues. One approach would be to assume that since the residue is not observed it should never occur in that position; this is clearly unrealistic, especially if the alignment contains only few sequences. Another is to use a substitution matrix to arrive at weighted average scores for all unobserved amino acids. However, this will reduce the specificity of the PSSM if indeed the variability in a motif block, that could be observed if the entire sequence family was known, is well represented by the alignment.

A statistical approach to this problem of unobserved amino acids adds imaginary pseudo counts to the observed counts of each amino acid at a position, based on some belief about amino acids expected to occur there. During a search, the query sequence is aligned with each block at all possible positions, and an alignment score is computed by adding the scores for each position. Before search scores are used to rank the blocks relative to each other, the blocks are calibrated to provide standardized scores that are comparable between blocks regardless of varying block widths and number of aligned sequences. The statistical significance of search results is reported as E-values.

5.2.3.3 *Full domain alignment methods*

5.2.3.3.1 *Profiles*

Profile analysis is used to perform whole domain alignment of a family-based scoring matrix against query sequences by dynamic programming. Profiles maximize information by including all positions in a multiple domain alignment; they allow partial matches and can detect distant relationships with only few well-conserved residues. A profile is a weight matrix comprising two components for each position in the alignment: scores for all 20 amino acids and variable gap opening and extension penalties. Specifically, the M_{kj} element of the profile is the score for the *j*th amino acid at the *k*th position of the alignment. A profile can be seen as a mathematical model for a group of protein sequences. This model is more complex than a simple weight-matrix model in that it contains position-specific information on insertions and deletions in the sequence family, and is quite closely related to the one employed in Hidden Markov Models for sequences. However, in contrast to Hidden Markov Models, profiles allow the generation of models with good discriminatory power from a small number of sequences.

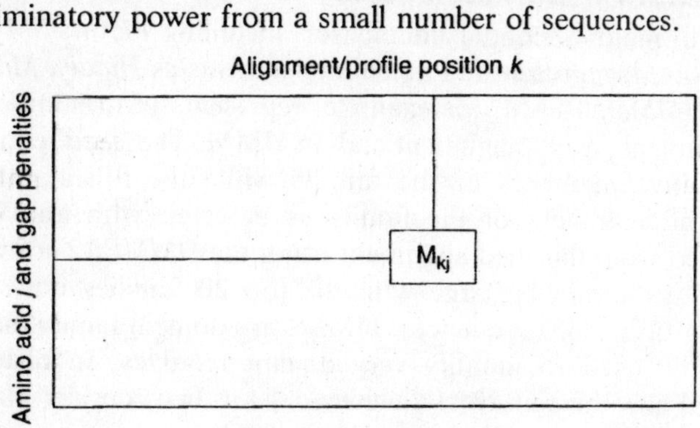

Figure 5.3. Profile matrix.

As alignments often consist of many close! related sequences together with a few much more divergent ones, sequence weighting schemes which upweight divergent sequences while downweighting closely related group can be used to improve profile sensitivity. Earlier profile methods created the amino acid substitution scores by summing Dayhoff *exchange matrix* values according to the observed amino acids in each column of the alignment. Other residue substitution matrices, particularly the *BLOSUM* and *Gonnet series,* have subsequently been found to improve the signal-to-noise ratio in profiles. BLOSUM 45, a moderate to high divergence matrix, works well as a starting point. Ideally, several matrices should be tested during profile construction to reach an appropriate balance between specificity (ability to detect true positives and exclude false positives) and sensitivity (ability to detect divergent family members as true positives while still keeping the expected number of false positives within very low bounds).

As in all sequence searches, the length of the query affects the tolerance of noise introduced by high divergence matrices. Small domains need more stringent matrices such as BLOSUM62 (or Gonnet PAM120/PAM160), whereas profiles for larger domains are often more sensitive at higher divergence (Gonnet PAM250-PAM350). Gap penalties are reduced at positions with gaps, according to the length of the longest insertion spanning that point in the alignment. As yet there is no satisfactory way of calculating these penalties. The user should conduct several trials varying the gap parameters in order to optimize alignment of the query sequence with the profile and the detection signal to-noise ratio.

5.2.3.4 Hidden Markov Models

Several major secondary databases, including *PFAM, SMART,* and *TIGRFAM,* represent full domain alignments as *Hidden Markov Models (HMMs).* PFAM, for example, represents each family as a 'seed alignment', a full alignment and an HMM. The 'seed' contains representative members of the family, while the full alignment contains all members of the family as detected with an HMM constructed from the seed alignment using the *HMMER2* software. Full alignments can be large with the top 20 families now each containing over 2500 sequences. HMMs are domain family models that can be used to identify very distant relatives. In order to understand how this sensitivity is achieved, let us first consider Markov processes in a more general way.

A *Markov process* is a process that can be decomposed into a succession of discrete states. The simplest Markov process is a first order process, where the choice of state is made purely on the basis of the previous state. The behaviour of traffic lights is a first-order Markov process, it consists of a set sequence of signals (red-red/amber-green-amber) and the transition to the next state only depends on the present colour of the signal. However, various examples exist where the process states are not directly observable. In speech recognition, the sound we hear depends on such factors as the action of the vocal chords, the size of the throat, and the position of the tongue. The detectable sounds are generated by the internal physical changes in the person speaking; and the observed sequence of states (sounds) is somehow related to the hidden process states (physical changes).

In cases like this, we can define an HMM, a statistical model of an underlying (hidden) Markov process which is probabilistically related to a set of observable state changes. *Profile HMMs* are bioinformatics applications of HMM modeling that represent protein domain families. The discrete states of a profile HMM correspond to the successive columns of a protein multiple sequence alignment: A *match state* (M) models the distribution of residues allowed in the column. An *insert* (I) and a *delete* (D) *state* at each column allow for insertion of one or more residues between that column and the next, or for creating a gap. States have associated *symbol emission probability distributions* and are connected by state transitions with associated *transition state probabilities.*

One may therefore conceptualize profile HMMs as process state models that generate (observable) sequences; they describe a probability distribution over a potentially infinite number of sequences.

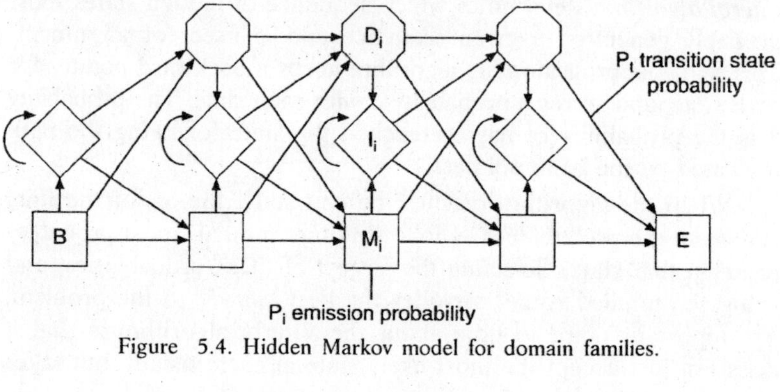

Figure 5.4. Hidden Markov model for domain families.

Each time a complete path is traversed from the begin (B) to the end (E) state, a hidden state sequence is created together with an observed (gapped) amino acid residue sequence. A profile HMM is similar to a profile in that each column in the profile can be seen as a match state, and the values in the column as emission probabilities for each of the 20 possible amino acids. The position-specific gap weights represent transition probabilities for moving to an insert or delete state from a match state.

The main difference between profile and HMM models is that the profile model requires that the transition from a match state to an insert state or a delete state have the same probability. This reflects the observation that an insertion in one sequence can be viewed as a deletion in another. In principle, HMMs can be developed from unaligned sequences by successive rounds of optimization, but in practice, protein profile HMMs are built from curated multiple sequence alignments.

In the latter case, parameter estimation is simplified and observed counts of symbol emissions and state transitions can be directly converted into probabilities. These probability parameters are then usually converted to additive log odds scores before aligning and scoring a sequence (p_x, probability of residue x emitted at match state; f_x, expected back ground frequency of x in sequence database; log p_x/f_x score for residue x at this match state). Once an HMM is generated, standard dynamic programming algorithms can be used for aligning and scoring sequences with the model.

HMM searches resemble later round PSI-BLAST searches, with position-specific scoring for each of the states (amino acid match state, insertion, and deletion) over the length of the sequence. The *Viterbi algorithm* determines which sequence of hidden states most probably generated a given sequence and is used for alignment. This dynamic programming algorithm keeps a backward pointer for each state, and stores a probability δ with each state. The probability d is the probability of having reached the state following the path indicated by the back pointers.

When the algorithm reaches the end state, the δs for the final states are the probabilities of following the optimal (most probable) route to that state. Selecting the largest of these probabilities, and using the implied route, provides the best answer to the problem. An important point to note about the Viterbi algorithm is that it does not just accept the most likely state at each instant, but takes a decision based on the whole sequence. It will therefore not matter if there is a particularly unlikely event somewhere in the sequence, provided the context is significant. Using the example of speech recognition once more, an intermediate sound may be indistinct or lost, but the overall sense of the spoken word may be detectable.

The same principle applies to matching distantly related protein sequences with highly diverged positions to a family HMM. The *forward algorithm* assigns a probability for the generation of a specific sequence by a given HMM and is used for scoring sequences. One method of calculating the probability of the observed sequence would be to find each possible sequence of hidden states that would generate it, and sum these probabilities. Calculating the probability in this manner is computationally expensive, particularly with large models or long sequences.

In order to reduce the complexity of the problem, it is possible to calculate the probability of reaching an intermediate state in the HMM as the sum of all possible paths to that state. The sum of all partial probabilities is the sum of all possible paths through the model, and hence is the probability of observing the sequence given the HMM. The forward algorithm finds this probability exploiting recursion in the calculations to avoid the necessity for exhaustive calculation of all paths. Using this algorithm, it is straightforward to determine which of a number of HMMs best describes a given sequence—the forward algorithm is evaluated for each, and that giving the highest probability is selected.

5.2.3.5 Statistical significance of profile and HMM hits

When searching profile or HMM databases with a query sequence, every match receives a numerical *raw score* generated by the search algorithm applied. Usually, only matches with scores above a certain threshold are reported, and every profile or HMMs applies its own score thresholds. To facilitate the interpretation of match scores, they are mapped to a new scale, which possesses a well-defined statistical meaning. Two such scales are the *E-value* for HMMI and the *normalized score* for PROSITE profiles and the Gribskov profile collection. Normalized scores have an inverse log relationship to the E-value: A normalized score of 10 corresponds to an E-value of around 1×10^{-2} and 10.5 to an E-value of around 2×10^{-3}. Any normalized score above 11 indicates a very significant match.

5.2.4 Secondary Database Searching: A Worked Example

The *InterPro* database allows efficient searching, as it is an integrated annotation resource for protein families, domains and functional sites that amalgamates the efforts of the *PROSITE*, *PRINTS*, *Pfam*, *ProDom*, *SMART* and *TIGRFAMs* secondary database projects. Release 4.0 of InterPro contains 4691 entries, representing 3532 families, 1068 domains, 74 repeats and 15 sites of post-translational modification. Overall there are 2,141,621 InterPro hits from 586,124 SWISS-PROT + TrEMBL protein sequences. Search results are reported in PFAM as a table of significant hits showing bit scores and E-values, and also in graphical format depicting the positions of matched domains along the query sequence.

PFAM also allows one to retrieve a graphics table of other proteins known to share a given domain with the query, as shown by three example proteins sharing the kringle domain with human TPA. The human apolipoprotein *A* provides a dramatic example of *domain duplication*—it contains 38 kringle domains. Interpro reports search results in a table that groups together hits to the same domain obtained by searching its constituent secondary databases. The graphical display allows one to clearly distinguish *short single motifs* matched by PROSITE regular expressions (PSxxxxx), *multiple motifs* matched by PRINTS fingerprints (PRxxxxx), and *whole domains* matched by HMMs from PFAM (PFxxxxx) and SMART (SMxxxxx), PRO SITE profiles (PSxxxxx) and consensus sequences from PRODOM (PDxxxxx). Clicking the hyperlinked domain boxes enables the retrieval of more detailed information on matched domains.

5.3 CONTEXT-DEPENDENCE OF PROTEIN FUNCTION

Sequence analysis techniques provide important tools for the prediction of biochemical function, but show clear limitations in predicting a protein's context-dependent functions and role in one or more biological processes. Sequence analysis techniques are complemented and extended by function prediction from structure and data from transcriptomics and proteomics which can enhance understanding of gene and protein function in complex biological networks.

6

Data Management

6.1 ANALYZING DATA FROM 2D-PAGE GEL

6.1.1 Raw Data from 2D-PAGE Gels

Two-dimensional polycrylamide gel electrophoresis (*2D-PAGE*) is a method used to separate proteins according to charge and mass. The resolution of the technique is such that thousands of proteins can be distinguished, providing a diagnostic *protein fingerprint* of any particular sample. After the gel has been run, it is stained to reveal the position of individual proteins. These appear as spots of varying size, shape and intensity.

The role of bioinformatics in protein gel analysis is the extraction of useful information from the positions and intensities of the protein spots. 2D-PAGE data can be used to derive general information on protein expression profiles without any knowledge of which specific proteins are actually present on the gel. However, the most powerful approach to 2D-P AGE experiments is to couple the expression analysis with protein annotation by mass spectrometry.

6.1.2 Data Processing

Data are extracted from 2D-PAGE gels in several stages. First, the stained protein gel is scanned to obtain a digital image. Individual protein spots are then detected and quantified, and the intensity of the signal for each spot corrected for local background. The quality of the digital image is important since its *spatial resolution* determines how accurately protein spot sizes are recorded and its *densitometric resolution* determines how accurately the intensity of each spot is recorded. Protein spot measurement involves a special set of problems that are not found with microarray data. The features on a microarray are arranged in a precise grid and the signals tend to be regular, discrete and nonoverlapping. Conversely the signals in a protein gel are irregularly distributed, the sports vary widely in morphology, and it is often the case that spots join together in clumps or lines that are difficult to resolve.

Several algorithms are available which address these issues and generally these are based on either *Gaussian fitting* or *Laplacian of Gaussian* (*LOG*) *spot detection*. Spots whose morphology deviates from a single Gaussian shape can thus be interpreted using a model of overlapping shapes. A simpler approach is *line and chain analysis*, in which columns of pixels from the digital image are scanned for peaks in signal density. This process is repeated for adjacent pixel columns allowing the algorithm to identify the centers of spots and their overall signal intensity (*signal volume*).

A further approach is known as *watershed transformation* (*WST*). In this method, pixel intensities are viewed as a topographical map so that hills and valleys can be identified. This is useful for separating clusters, chains and shouldered spots (small spots overlapping with larger ones) and also for merging regions of a single spot. The output of each method is a *spot list*, in which each individual protein spot is identified by the *x* and *y* coordinates of its center.

6.1.3 Gel Matching

An important application of 2D-PAGE is the analysis of *differential protein expression*. This can be used, for example, to look for proteins that are induced or repressed by particular treatments or drugs, to look for proteins associated with disease states, or to look at changes in protein expression during development. Differential protein expression can only be studied by running a series of 2D-PAGE gels with alternative samples, and searching for novel spots or spots whose intensity changes significantly in different gels. Note that any observed changes, for example the presence of a novel protein spot, may not reflect the *synthesis* of a new protein.

Instead, the new sport may reflect the *post-translational modification* of an existing protein, for example phosphorylation or glycosylation. Such modifications can radically alter the mass or charge of a protein, leading to different migration behaviour during electrophoresis. Due to minute variations and inconsistencies in the

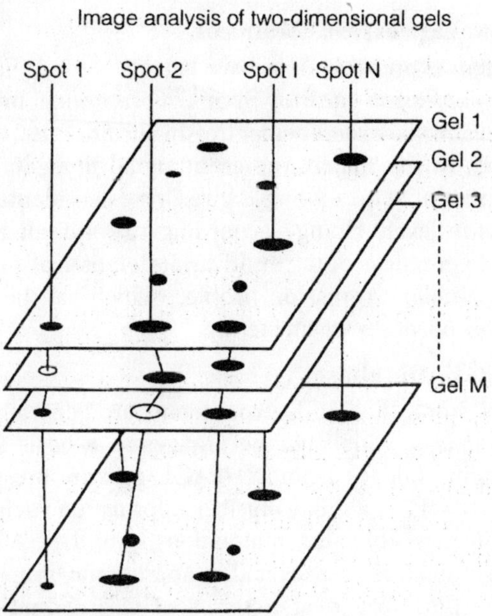

Figure 6.1. Principle of gel matching. Individual spots are compared and matched across a series of 2D-PAGE gels.

chemical and physical properties of electrophoretic gels, it is impossible to exactly reproduce the conditions of anyone electrophoresis experiment. This means that, even in a series of gels using exactly the same sample and electrophoresis parameters, the positions of individual protein spots are never the same.

In order to compare serial gels and identify novel protein spots or spots that vary in intensity, it is necessary to use *gel matching*. Generally this involves establishing the positions of several unambiguous *landmark spots* and then using algorithms to match the positions of the remaining spots. To bring the spots from two gels into register, simple image manipulations such as stretching and rotating can be carried out. This may be assisted by incorporating spot intensities and the (physical) distances between neighbouring spots as variables in the algorithm. For example, an approach known as *propagation* involves determining the distance between a given landmark spot and all neighbours. If matches are found on other gels, the neighbouring spots can be used as new landmarks, and the process reiterated. An implementation of such algorithms is available as the program *MELANIE II*.

6.1.4 Protein Expression Matrices

Once protein expression data have been recorded they are built into a *protein expression matrix*. Spots, representing proteins, are arranged in rows, while experimental treatments are listed in columns. The data points in the matrix represent signal intensities for each spot. Comparison of values along a given row can identify proteins whose expression levels change according to different treatments. As with gene expression data, multivariate statistical analysis can group data by similar expression profiles, which can be used both for classification purposes or clustering.

6.1.5 2D-PAGE Databases

The results from 2D-PAGE experiments are generally stored in *2D-PAGE databases*, which use gel images as a basis for protein annotation. Digital images from 2D-PAGE gels are presented, and each protein spot acts as a link to further information such as protein name, molecular mass, pI value, annotations from the SWISS-PROT database, bibliographic references and, if appropriate, graphs showing how spot intensities vary over a series of gels. Some 2D-PAGE databases are free standing and may come packaged with gel analysis software. For example, the analysis suite *PDQUEST* incorporates its own database, which may be installed locally and used to store the 2D-PAGE experimental data from a single laboratory.

There are also free software packages available that can be used to set up 2D-PAGE databases, such as *Make2ddb*. The construction of 2D-PAGE databases on the World Wide Web (WWW) allows 2D-PAGE data to be shared and compared over the Internet. There are over, 30 such databases currently available, many showing 2D-PAGE data either for specific organisms or specific systems. For example, the HEART-2DPAGE contains 2D-PAGE data related to heart development, physiology and disease.

All these pages are indexed by the ExPASy server on the *WORLD-2DPAGE*. The format of data presentation is standardized. Images of 2D-PAGE gels are presented overlain with a grid showing pI values and molecular masses. This allows each protein spot to be identified on the basis of its physical and chemical properties. Individual spots can be clicked, linking the user to an annotation page. On many gels, most of the spots are not annotated, and those with annotations are highlighted in some manner. For example, on the SWISS-2DE database each annotated protein spot is marked with a red cross. Annotations can be viewed as an original SWISS-PROT file.

Most of the fields are self-explanatory but note that the 2D field may include data on both 2D-PAGE electrophoresis and peptide mass fingerprinting where such experiments have been carried out. This is a catalog of some of the 2D gel images found on the WWW, with associated search facilities so that images can be abstracted by key words. 2DWG represents a very good introduction to the world of 2D gels. The images are stored in formats such as GIF that Internet browsers can handle and there is an associated Java applet, called *Flicker*, to facilitate the comparison of two gels. The basis of flickering is that the images of two gels can be rapidly alternated to identify matching spots. Another useful applet is *CAROL* which uses point pattern matching to compare any two gel images over the Internet.

6.2 ANALYZING PROTEIN MASS SPECTROMETRY DATA

6.2.1 Raw Data from Mass Spectrometry

Mass spectrometry (*MS*) is a method for accurately determining the *mass/charge ratio* (*m/z*) of ions in a vacuum, thus allowing the precise determination of *molecular masses*. These raw data can be used in three different approaches for the identification of proteins. In *peptide-mass fingerprinting*, a protein is digested with a specific *cleavage agent* (usually the enzyme trypsin) and the masses of each of the resulting peptides are determined. These masses are then used in *correlative database searching* to identify the protein. In *fragment ion searching* peptide fragments are generated as above and then fragmented in a *collision cell* between two quadrupole mass analyzers (this is called *tandem MS*, often abbreviated to *MS/MS*) or by a process known as *post-source decay*, which occurs when matrix-assisted laser desorption/ ionization (MALDI) MS is used at a higher acceleration voltage than usual.

The resulting *fragment ions* are short peptide fragments. The molecular masses of such fragments can be used to search not only protein databases, but also other sequence repositories including dbEST. Both these approaches require that the protein in question has been identified and its sequence deposited. Where this is not the case, *de novo sequencing* can be carried out. In this method, peptide ladders are generated either by sequential chemical degradation of terminal amino acid residues or by separating the fragment ions generated as above into nested sets. Mass differences between sequential fragments correspond to the known masses of individual amino acids, thus allowing protein sequences to be deduced without any correlative information from sequence databases.

6.2.2 Virtual Digests

It would be impossible to link peptide mass data to proteins in sequence databases such as SWISS-PROT without first knowing the *expected* peptides from such proteins. This information can be obtained by carrying out *virtual digests,* that is, theoretical digests based on the known protein sequence and the known specificity of the cleavage agent used. Cleavage agents with high specificity are most suitable and the endoproteinase *trypsin*, which cleaves a polypeptide chain after each basic amino acid (lysine or arginine) providing the next residue is not a proline, is the most widely used. Given a protein of known sequence, the *tryptic peptides* can be predicted. For example, a protein with the sequence shown below

MCLTAKGAATCSATFRYLIFALSLATKPACALLASALLARACATTAVA

would generate the following tryptic peptides:

MCLTAK GAATCSATFR YLIFALSLATKPACALLASALLAR ACATTAVA.

The four peptides provide four theoretical molecular masses. Correlation between these theoretical masses and the actual masses obtained in a mass spectrometry experiment would provide very convincing evidence for a database match. Of course the same molecular masses could be obtained in many other ways.

Each of the four peptides has a potentially very large number of anagrams (same amino acids in a different order), which would all have the same molecular mass.

Theoretically, this places limitations on the usefulness of the technique but in practical terms the chances of a series of peptides from the same protein generating spurious matches to the same deposited protein sequence because of permutations in the order of amino acids are very small indeed. Generally, correlation between two or more peptides is taken to be unambiguous confirmation of a database match. Another theoretical limitation is that the amino acids leucine and isoleucine have the same molecular masses. This provides a small technical problem in *de novo* sequencing, but for peptide mass fingerprinting and fragment ion fingerprinting it does not have a practical impact.

6.2.3 Dual Digests

Peptide mass fingerprinting allows the rapid identification of proteins if they are already represented in databases such as SWISS-PROT. Where this is not the case, both peptide mass fingerprinting and fragment ion searching can be used to match mass spectrometry data to other sequence databases, including the expressed sequence tag (EST) database dbEST. The problem with this approach is that the databases contain a large amount of irrelevant data (e.g. noncoding sequence and other 'noise'), which can reduce the efficiency of the search. More confidence can be placed in any results if *dual digests* are carried out (i.e. combining the information from two protease digests using enzymes with different specificities, such as trypsin and endoproteinase LysC).

Another approach is to carry out a single digest with the protein in the native state and then carry out the same digest after modifying the protein, for example by methylation. Furthermore, since lysine (K) and arginine (R) are two of the most common amino acids in proteins, there is a relatively large number of doublets and triplets (e.g. RR, KKR, KRK). Trypsin cleaves randomly at such sites generating peptide fragments with *ragged termini*. More confidence can be placed in database hits that are compatible with such peptides because this is effectively the same as searching with a larger number of peptides.

6.2.4 Database Search Tools

A number of algorithms have been developed for sequence database searching using MS data. Among these, the most commonly used is *SEQUEST*, which works by searching for all peptides in the specified database(s) with the same mass as a given peptide ion. Then, a virtual digest is performed on the matched protein and a theoretical mass spectrum generated. The data from the theoretical mass spectrum are then compared to the experimental data and the best matches are scored. In a different approach, peptide mass data are used to generate a collection of possible sequences, and this profile is used as a query in a modified BLAST or FASTA search. A program called *Lutkefisk* has been developed for this purpose.

6.2.5 Limitations of MS Analysis

Although a powerful technique for protein annotation and sequencing, there are some limitations to MS, which need to be taken into account when interpreting and analyzing experimental data. One of the most important factors to take into account is that MS data may not match any database entry due to the presence of an unknown *post-translational modification*. Where such modifications are known, exact mass differences between unmodified and modified amino acids can be predicted. Indeed several algorithms, including SEQUEST, have built-in parameters for detecting such modifications. However, the presence of a modified residue should always be confirmed experimentally. Another potential problem is the occurrence of nonspecific proteolysis. This depends on the purity of the cleavage agent used.

Many algorithms will carry out peptide mass searches without a specified cleavage agent to take nonspecific proteolysis into consideration. A common problem is that protein spots isolated from 2D-PAGE gels often contain a mixture of proteins, and these contaminants may be difficult to identify. Finally, imperfect matches may result because the actual protein does not exist in the database, but a close homolog from the same species or a different species, which may have a related sequence, does exist. This is often the case if a protein contains a single nucleotide polymorphism leading to two or more variants. Programs such as MS-BLAST and CID entify use sequence candidates from tandem mass spectrometry data as the input for a homology search.

7

Storing Data

Bioinformatics methods aim to extract information about biological functions and mechanisms from computer-held biological data. These data range from genomic sequence data to the transcriptomics and proteomics data discussed in previous chapters. Data management techniques enable these increasingly huge amounts of data to be stored and extracted from computers reliably and efficiently. Most of these techniques were developed for business applications, where the data are structured and it is clear what object any particular piece of data relates to, such as in the case of recording a customer's name, address and account balance. However, managing biological data and extracting useful information from them poses particular problems for the following reasons.

1. Biological data come in many different forms: structured, as in GenBank or PDB file formats; unstructured, as in biomedical literature abstracts; sequence, as in genomic or protein sequences; or more complex, as in biochemical pathway data.
2. There is no single biological object to which the data refer: individual users of the data have their own perspective depending on the science they are carrying out, ranging from the atomic level to that of whole organisms.
3. Similarly, the biological functions and mechanisms to which data refer are also dependent on the perspective of the user of the data.
4. Even if the appropriate biological object can be identified, there are many incompatible formats and nomenclatures for the data about that object and its functions.

5. The data about any given biological object are likely to be held at many different sites,each with a different approach to coping with the previous problems.

7.1 DATA MANAGEMENT CONCEPTS

7.1.1 Databases and Database Software

The term *database* can be used simply to mean any collection of data stored in a computer, perhaps even in a single file. Biological data have indeed frequently been held in simple files often with tags to indicate the meaning of a particular piece of data as, for example, in the case of a GenBank file. Files may also have a fixed record format, as in the case of a PDB file. As discussed below, problems often arise with file-based databases, and this has led to the development of database management techniques to overcome such problems.

Using these techniques, a database is still a collection of data stored in a computer, but the storage of and access to the data is controlled by specialized software called a *database management system* (DBMS). In practice, even biological data which have traditionally been in a file format, such as GenBank or PDB data, are in reality now managed by a DBMS and the familiar file formats are retained as one way of making such data available to scientists.

7.1.2 Why are DBMSs Useful?

Scientists have traditionally written programs in languages such as Perl, C or C++ to extract data relevant to their work from files, perform processing to generate new data of interest, and then write those data to a new file. This way of working leads to the following problems.

1. Data are not shared. The data generated by each scientist are stored in different files, the format of which will be dependent on the precise programs written by the scientist. It will not be possible for other scientists to share those data without knowing of the existence of the files and their format.
2. Data are not integrated. The new files generated by each scientist have no relationship to the files of data from which they were derived, nor with the many files being generated by other scientists. The information which would be available if the data stored in the separate files were integrated and inter-related is lost.

3. Data are duplicated. The same or closely related data are generated many times by different programs, and stored in different files. This inevitably leads to problems when revisions are made to one file of generated data, perhaps to correct an error, but the other files of identical or closely related data are not revised. Multiple versions of data which should be consistent are then inconsistent.
4. Programs are dependent on file formats. If the format of a file is changed, perhaps to add some new type of data, then the programs which access that file have to be modified to take account of the new format.

DBMSs were developed to overcome these problems. They do this in the following ways.

1. Data are held in a single shared resource: the database. Each scientist still writes programs to access data, but the format of the data is defined centrally for the database rather than being determined by an individual scientist's program. Further, if the data generated by the scientist are themselves stored in the database, they can be shared by other scientists.
2. Data are integrated in the database. When data are stored in a DBMS-managed database, the DBMS enables the relationships between all the data to be explicitly represented, and the information inherent in those relationships is available.
3. Data need not be duplicated. In reality data *are* almost always duplicated in a DBMS-managed database, but the duplication is controlled by the DBMS. The reasons for this duplication are to enhance performance, and to enable recovery of lost data if some failure results in their loss.
4. Programs are independent of file formats. The format of the data may be changed or new types of data added to a DBMS-managed database, but existing programs which have no requirement to access the new types of data need not be modified: the DBMS itself ensures existing programs still work as though there had been no change to the data formats.

The difference between file-based and DBMS-managed data is illustrated in Figure 7.1. The upper half of the figure shows file-based 'databases': many different programs access data in multiple files and write out new data to yet more files. It becomes an impossible task to maintain data as a shared, integrated, consistent resource. The lower half of the figure shows DBMS-managed data.

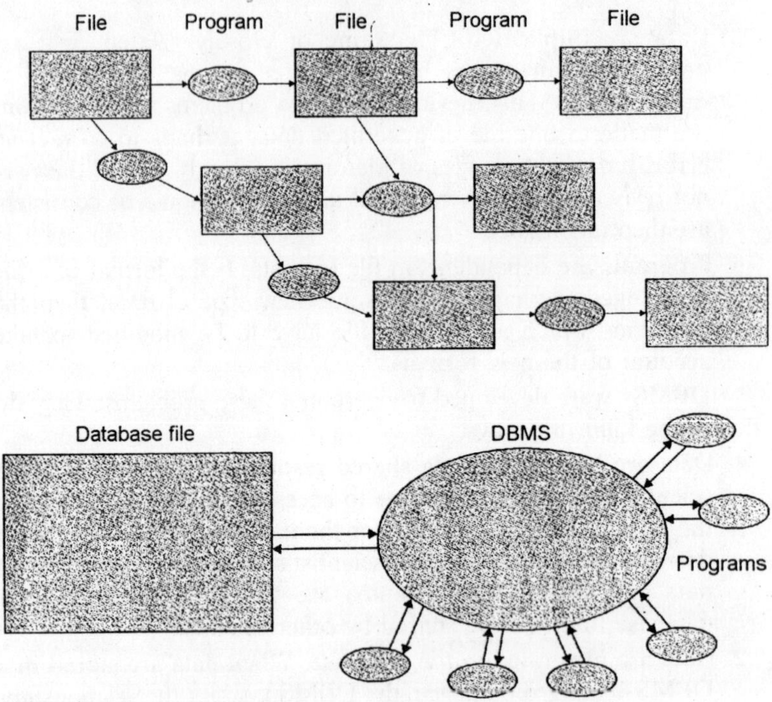

Figure 7.1. File-based v DBMS-based data.

Programs now access data in a shared, integrated database by using the services of a DBMS, which is able to ensure data is shared, integrated and consistent.

It might be imagined from the above that having data managed by a DBMS rather than using ordinary files is always advantageous. Indeed, many scientists use a DBMS as though the very fact of having data in a 'real' database gives benefits. In fact there are disadvantages in using a DBMS including the following.

1. A DBMS is expensive. This can include the cost of purchasing the DBMS software, but equally importantly a DBMS is a large piece of software which requires more computer resources such as memory than conventional programs accessing files. File-based access to data does not incur these costs.
2. DBMSs are complex. If a scientist writing a Perl program to access a file finds the program is not working as expected, there is every chance that the source of the problem can be discovered and corrected. If the same program is using a DBMS,

there are many areas of the DBMS's operation which are complex and can only be corrected with specialist knowledge of the DBMS.

3. Access to data by a DBMS may be slow. The DBMS software is potentially carrying out a great deal of work to enable access to data reliably by many different programs concurrently and there is an inevitable overhead associated with this. A skilled programmer will always be able to write a program accessing file-based data more quickly.

In summary, the advantages of a DBMS are that it provides shared, integrated access to data, with the ability to modify the types and formats of stored data for the needs of new applications, while allowing existing applications to continue working without modification. Usually these are significant advantages to a scientist with data, but if the costs arising cannot be justified, then it makes sense to continue working with file-based data.

7.1.3 Relational and other Databases

Another problem with file-based data is that it can be difficult to understand what information is contained in the file. A PDB file, for example, certainly contains much information about the 3D structure of a macromolecule, but it is no simple matter to understand from the file what are the biological and other entities and their relationships inherent in the data. An important feature of a DBMS is that it supports a *data model* which makes it easier to understand what information the database contains. Many different types of data model have been developed since the 1960s, but for over 10 years the *relational model* has been dominant, and DBMSs supporting the relational model are currently the most widely found.

The essential characteristic of the relational model is that all data are represented as being stored in tables. Some columns from two tables, PDB_DOM and PDB_DOM_SEG are shown with some example rows in each. PDB_DOM records the CATH structural classification for each identified domain in the PDB. A domain may encompass non-contiguous sections of a chain, each such section being known as a domain segment. PDB_DOM_SEG records the start and finish residue numbers of each domain segment. Once the decision has been made to use a relational database, there are a large number of different relational DBMSs which are available. These range from relatively simple products such as *MySQL* through products with more features such as *PostgreSQL,* both of which

PDB_DOM

PDBCODE	CHAIN	DOMAIN	C	A	T	H	S35	S95	S100
1lga	A	1	1	10	520	10	1	2	1
1lga	A	2	1	10	420	10	2	2	1
1lga	B	1	1	10	520	10	1	2	2
1lga	B	2	1	10	420	10	2	2	2

PDB_DOM_SEG

PDEBCODE	CHAIN	DOMAIN	SEGMENT	RESNUM1	RESNUM2
1lga	A	1	1	11	144
1lga	A	1	2	269	299
1lga	A	2	1	145	268
1lga	A	2	2	300	339
1lga	B	1	1	11	144
1lga	B	1	2	269	296
1lga	B	2	1	145	268
1lga	B	2	2	297	339

Figure 7.2. Example relational table data.

may be obtained freely over the World Wide Web, to more complex products such as *Oracle* which are intended for commercial use and can be expensive to buy. These products all have the essential characteristic of a relational DBMS in that the representation of data is in the form of tables; where they differ is in the sophistication of their features to enable the management of large amounts of data accessed concurrently by many users securely, reliably and efficiently.

While a financial institution would need the levels of security and reliability supported by a product such as Oracle, for many scientists the features supported by free software such as MySQL and PostgreSQL are entirely adequate for their needs. Indeed, the simpler products, without the overhead of features found in more complex products, may have performance advantages for straightforward access to data. Further reading about DBMS products can be found in the books by Yarger *et al.* (MySQL), Worsley and Drake (PostgreSQL) and Abbey *et al.* (Oracle). The relational model has the great advantage of being a very simple way of representing data, but when there are complex relationships between entities, as is the case for many biological data, it can be difficult to understand how the tables represent those relationships. Each PDB structure may have many domains, and each PDB structure may have many chains.

The relationship between domains and chains is a many-many one: namely, each domain may be related to many chains, while each chain may be related to many domains. However, the nature of these relationships is not clearly represented in the tables shown. It is desirable that any data model makes it easy to understand what relationships are represented therein, and the relational model does not always achieve this. An alternative data model, the *object-oriented model,* is supported by some DBMSs. This represents objects and the relationships between objects directly, as well as controlling the way in which objects may be created and accessed in the database.

So, for example, rather than having data about a protein domain spread across many tables as would be the case in a relational database, in an object-oriented database all the information about a protein domain, including what a domain is related to and the ways in which domain data may be created and accessed, would be brought together in the definition of a protein domain object. Some DBMSs have been designed specifically for managing scientific data. One having features specifically for managing genetic map, physical map and sequence data is *AceDB,* originally designed for the C. *elegans* sequencing project.

AceDB uses an object-oriented model with each object having a tree-like hierarchical structure. Graphical interfaces for displaying the map and sequence data come with the DBMS. Although an object-oriented model can be well-suited to representing data with complex relationships, the simplicity of the relational model and the strengths of well-established relational DBMSs in supporting reliable, shared access to integrated data have ensured their continued wide use.

7.2 DATA MANAGEMENT TECHNIQUES

Whatever relational DBMS product is used, the fundamental requirement of any user is knowing how to access the data in the database. Those scientists developing their own data bases additionally need to know how to design and create a relational database. Both of these tasks require an understanding of how a relational DBMS works, in order that the database may be used in an efficient way. The techniques underpinning such use of a relational database are now examined.

7.2.1 Accessing a Database

The standard language for accessing a relational database is *SQL* (Structured Query *Language*): no matter which relational DBMS product is used, SQL will be supported. In fact SQL is more than just a language to access the data in a relational database, it is also the language used to create the tables in a database and insert, modify or delete rows of data in those tables. SQL has a SELECT statement to retrieve rows from tables. For example

```
SELECT C, A, T, H
FROM PDB_DOM
WHERE PDBCODE = `1lga'
AND CHAIN = `B'
```

This query has three parts or clauses, a SELECT clause, a FROM clause and a WHERE clause. Every SELECT statement has SELECT and FROM clauses, while the WHERE clause is optional. If this SELECT statement were executed with the example tables, the rows retrieved forming the result would be:

C	A	T	H
1	10	520	10
1	10	420	10

The SELECT clause specifies which values in particular appear in each row in the query result - in this case C, A, T and H values. The FROM clause of the query specifies that data are to be retrieved by accessing the PDB_DOM table. If the query had no WHERE clause, the C, A, T and H values from every row in PDB_DOM would appear in the result. If the query does have a WHERE clause, as in this example, only rows which satisfy the WHERE clause conditions produce values in the query result.

Normally queries will require access to data in more than one table. For example, suppose we wish to know the CA TH classification of the domain containing residue 100 of PDB structure llga chain A. A possible SELECT statement is:

```
SELECT C, A, T, H
FROM PDB_DOM, PDB_DOM_SEG
WHERE PDB_DOM_SEG.PDBCODE = '1lga'
AND PDB DOM SEG.CHAIN = 'A'
AND PDB DOM SEG.RESNUM1 <= 100
AND PDB DOM SEG.RESNUM2 >= 100
```

```
AND PDB_DOM.PDBCODE = PDB_DOM_SEG.PDBCODE
AND PDB_DOM.CHAIN = PDB_DOM_SEG.CHAIN
AND PDB_DOM.DOMAIN = PDB_DOM_SEG.DOMAIN
```

producing result:

C	A	T	H
1	10	520	10

The query specifies both the PDB_DOM and PDB_DOM_SEG tables in the FROM clause: the query requires access to data in both tables. The first four conditions in the WHERE clause specify the conditions that a row in PDB_DOM_SEG must satisfy in order to figure in the result. For any row that does satisfy these conditions, the final three conditions in the WHERE clause specify the conditions which a row in PDB_DOM must itself satisfy in order for its C, A, T, H values to appear in the result, namely it must match the PDBCODE, CHAIN and DOMAIN values in the PDB_DOM_SEG row. SQL SELECT statements have many other features not illustrated by this example, but these features do not extend the fundamental way in which SELECT statements work. In essence, any SQL SELECT statement will retrieve data from one or more tables by using:

- a SELECT clause specifying the particular values which we wish to retrieve,
- a FROM clause to specify the tables which must be accessed, and
- an optional WHERE clause to specify conditions restricting the rows which we wish to retrieve from tables, and the conditions which specify how rows in different tables must match each other.

While SQL statements enable data to be retrieved from tables, developing a full application will usually require programming in a language such as Perl, Java or C++ for all the non-database tasks of the application. Hence, mechanisms have been developed to enable such languages to use SQL to access a database.

The main difficulty in using SQL in an ordinary programming language is that an SQL SELECT statement potentially results in many rows of data being retrieved to form the result, but programming languages are usually designed to handle a single row or record at a time. This mismatch is solved by use of an *SQL cursor* which enables row-by-row retrieval of results so a program can process a row at a time. The details of how to use SQL from a programming language depend on the particular language used. For Perl, a programming interface called *DBI* is widely used, as described by Descartes and Bunce. This is database independent, meaning it will work with most DBMSs including PostgreSQL, MySQL and Oracle.

A similar database-independent programming interface *IDBC* is used from within Java programs; White *et al.* provide a thorough tutorial and reference for JDBC. For C or C++ programs, an interface supporting *embedded SQL* is often used, in which SQL statements are interspersed with the ordinary C or C++ program statements. Embedded SQL is typically more dependent on the particular DBMS used than either Perl DBI or Java JDBC.

7.2.2 Designing a Database

There are two distinct but related aspects to designing a database, *logical design* and *physical design*. The former is concerned with designing the table structures which users of a relational database will see. The latter is concerned with designing the actual storage structures which underlie those table structures. This raises a fundamental aspect of databases which many database users remain unaware of: namely, the data model presented by a DBMS—a series of tables in the case of a relational database— may differ enormously from the storage structures maintained in reality by the DBMS to hold the data.

7.2.2.1 Logical design

Considering logical design first, there are usually many possible table designs for any given collection of data. This has introduced a single numeric identifer DOMAIN_ID to represent the combination of PDBCODE, CHAIN and DOMAIN. A new table PDB_DOM_ID has been introduced to record the relationship between the new DOMAIN_ID and the existing identifiers. The process of database logical design is that of choosing between different possible table designs such as seen here. One advantage is that domains are identified by a single numeric identifier. Hence, in queries, only this single identifier needs to be matched when combining rows from more than one table, rather than the values of three columns as was needed for the original design. Conversely, a disadvantage is that the query would involve three tables rather than the two or one of the alternative designs, and queries involving matching rows in multiple tables—*joins*—can be very slow.

PDB_DOM_ID

DOMAIN_ID	PDBCODE	CHAIN	DOMAIN
1	1lga	A	1
2	1lga	A	2
3	1lga	B	1
4	1lga	B	2

PDB_DOM

DOMAIN_ID	C	A	T	H	S35	S95	S100
1	1	10	520	10	1	2	1
2	1	10	420	10	2	2	1
3	1	10	520	10	1	2	2
4	1	10	420	10	2	2	2

PDB_DOM-SEG

DOMAIN ID	SEGMENT	RESNUM1	RESNUM2
1	1	11	144
1	2	269	299
2	1	145	268
2	2	300	339
3	1	11	144
3	2	269	296
4	1	145	268
4	2	297	339

Figure 7.3. Table design 2.

7.2.2.2 Physical design

Turning to physical design, while a relational database appears to consist of tables, in reality the data are stored in a variety of storage structures designed to support quick retrieval of data. It is the aim of physical database design to choose the most appropriate storage structures for a database. Here we consider only the use of index storage structures. Indexes in databases work in a similar fashion to indexes in books. A reader wanting to find pages in a book related to a particular topic can simply work through the book page by page examining each until one is reached mentioning the topic of interest. For short books this may be acceptable, but for long books it is simply too time-consuming.

Similarly for a table in a database, if rows are sought containing a particular value or values, the DBMS can examine each row one-by-one, but this will be slow for large tables potentially holding millions of rows. Therefore, DBMSs support indexes on tables. An index identifies which rows contain particular values for one or more columns in a table.

1. An index on the combination of values of columns PDBCODE and CHAIN in PDB_DOM_SEG would enable a DBMS to retrieve rows very quickly in that table with particular values for the PDBCODE and CHAIN columns, as was needed for the query for the combination of values ('1lga', 'N).
2. An index on the combination of values of columns PDBCODE, CHAIN and DOMAIN in PDB_DOM would mean that once a row in PDB_DOM_SEG had been retrieved, the DBMS could very quickly retrieve matching rows in PDB_DOM with the same combination of PDBCODE, CHAIN and DOMAIN values.

The example illustrates the two main ways indexes are used by a relational DBMS: to quickly retrieve rows in a table with a given combination of values therein, and additionally to retrieve rows in one table matching rows in a second table as required in a join of the tables. Although indexes may give improved query performance, they do result in slower update performance since as values in table columns are updated, associated indexes must also be updated by the DBMS. This cost, as well as the space required to store the indexes, must be borne in mind when considering a physical design.

7.2.3 Overcoming Performance Problems

Once a database has been designed and loaded with data, it can be a shock to find that the SQL SELECT statements written to query the tables run very slowly indeed. While this might be expected for large tables containing millions of rows, even with tables containing thousands of rows it is very easy to write SQL queries which take many minutes and possibly hours before results are produced. Conversely, it can be a surprise that a SQL SELECT statement which does query tables containing millions of rows nonetheless produces results almost instantaneously.

1. First, examine the SQL query itself. Has it been written correctly to return the results you intended? In particular, check that the query does not refer to more tables than are necessary in the FROM clause, or that the WHERE clause is missing a join condition specifying how the values in two tables should match.
2. Second, make sure the physical design of the tables is appropriate. It was noted that indexes enable a relational DBMS

to retrieve rows quickly. Have the indexes been defined to support the problem query? Are there other physical storage structures supported by the DBMS which would help the query run faster?

3. Third, if the physical design seems to be appropriate, is the DBMS choosing a sensible strategy to retrieve rows? When a DBMS is given an SQL query to execute, it considers alternative ways to produce the required results. If, for example, it needs to do a join retrieving rows in one table matching rows in a second table, it has a number of options. It could simply retrieve the rows in the first table one by one, and check each against every row in the second table to find matches. Alternatively, it could use an index to find matching rows in the second table. Yet another alternative would be for it to sort both tables on the values being matched, and then scan each sorted table in parallel looking for matches. The DBMS is carrying out *query optimization* in choosing between the many options it has for producing the results of a query, and sometimes it does not make the best choice. Many DBMS products have tools to enable a user to view the strategy used by the DBMS to execute a query - the *execution plan*—and provide ways enabling a user to change the execution plan so that the DBMS executes the query more quickly. This might involve writing the SQL query in a slightly different way, or running a utility which updates the information the query optimizer uses to choose an execution plan, or changing one of the parameters controlling the detailed way the DBMS works. The details will be specific to each DBMS.

If none of these results in the query running quickly enough, more drastic solutions must be considered:

1. Should the computer be upgraded with more memory or processors? DBMS products are notoriously greedy in their use of computer resources.
2. Should the logical design of the database be changed? For example, should the design be changed? The problem query would then have to retrieve data from only one table rather than two, so removing the table join which could be causing the performance problems. The implications of changing a table design are considered further below.
3. If all else fails, then there is no alternative but to accept that it is not feasible to run the query against the database. In that

situation, one would look for a simpler, more efficient query to extract data from the database, and then manipulate that data outside the database using a conventional language such as Perl or C++.

The option of changing the logical design of a database was noted above as one way of removing performance problems with some queries. Of course, it is highly desirable that the most appropriate logical design is chosen in the light of expected queries before the database is created. A change to the logical design of an existing database is not a task to be undertaken lightly, since that may then force changes to be made to all the SQL queries in applications retrieving data from the original tables. DBMS products reduce this problem by allowing the definition of *view* tables, which look like the original tables but do not exist as tables stored in their own right. Instead, view tables are defined by SQL queries which are run by the DBMS when needed, to give the impression that the view tables are actually stored.

In this way existing application programs may be able to continue with SQL queries as originally written, but those queries now reference view tables giving the appearance of the original table design. Since what appears to be a single view table may in reality be produced by an SQL query joining a number of real tables, retrieving data from a view table may be slow given the joins involved. Some DBMS products support *materialized views* which, like ordinary view tables, are tables defined in terms of an SQL query, but in which the data for the resulting table are explicitly stored as a real table. As changes are made to the tables on which the materialized view is defined, the DBMS can then sometimes automatically make corresponding changes to the stored materialized view table.

7.2.4 Accessing Data from Remote Sites

It was noted that the data about any biological object are almost certainly held at many sites rather than in a single database. Hence, scientists typically need to be able to access data from many sites in order to carry out their work. Traditionally, this has meant downloading files from remote sites and writing programs to access those local copies of files, which has added to the problems of file-based data. With the advent of the World Wide Web, it has become common for web-based interfaces to be made available for browsing and searching data at remote sites, but in order for a scientist then to write programs to process such data, there has often still been no alternative other than to download the data to a local machine. DBMS-based databases have a number of features to make it easier for programs to access data held remotely.

1. First, DBMS software usually follows a *client-server* model. This means that the database and the DBMS software for managing it can be at one *server* site, while an application program and the DBMS software needed to enable that program to access the database can be at a separate *client* site. The database file is at a server site, each program may be at a different client site, and the DBMS software itself runs partly at the server site and partly at each client site.
2. A further level of flexibility is provided by a *distributed database.* This enables the database data to be spread across a number of sites, so what appears to a program to be a single database may in fact be a series of databases distributed across multiple server sites.

The techniques just described may be sufficient if all the remote data are held in the databases managed by a particular DBMS product. Even if the remote data are held in databases managed by different DBMS products, there is generally no problem in writing a program to access those data. Both the Perl DBI and Java JDBC are designed to work with multiple databases managed by different DBMS products. The Java JDBC interface can also be used from a client program without any DBMS software being installed on the client machine: the DBMS software needed is downloaded automatically from the server site. Another useful feature of Java JDBC is that it enables data held in files outside any database to be accessed nonetheless as though they were stored in relational database tables. The real problem with accessing remote data stored at multiple sites managed by different DBMS products or outside any database is that the data are no longer integrated and shared.

Each database is completely separate and so, once more, with each box being a database rather than an individual file. All the problems of that model then reappear. One approach to dealing with such problems is to build specialized software to cross-link data in different biological data sources, together with facilities to enable data from those sources to be retrieved in response to a query. *SRS* (Sequence Retrieval System) was developed at EMBL to achieve this for molecular biology data, and provides facilities to link sequence data (nucleic acid, EST and protein) together with wider data sources such as protein structure and bibliographic data resources.

SRS supports querying of those resources from both Web-based interfaces and a command line interface suitable for use from programs. Some general technologies have been developed to enable data from multiple remote sites nonetheless to be treated as a single, integrated resource. In particular:

1. *Data warehouses* are formed by extracting data from primary data sources and then holding the data in a database specifically designed to allow the relationships between the different data from the primary sources to be modeled and searched.
2. *Mediator* software is designed to run between the server software of remote data server sites and the application programs of client sites. A client program is given the illusion that the data are all stored according to a single common data model. This is achieved by the mediator software intercepting client requests for data, translating those requests into the format required by a particular data server, and then transforming the data retrieved from a server site into a form consistent with the common data model for transmission back to the client program. The term *federated database* is sometimes used to refer to a system consisting of a number of server sites managing independent databases controlled by different DBMS products, but for which a single common data model is also supported to enable the databases to be accessed as a single resource.
3. *XML technology* enables data from a remote data server to be retrieved with tags which explicitly represent the meaning of the data in a form suitable for processing by computer programs. XML has become particularly important for enabling data to be passed between computer programs over the World Wide Web.

In recent years, DBMS products have become increasingly sophisticated in their support for these kinds of technologies. A DBMS no longer just manages data in tables in a single local database; increasingly it provides access to data stored locally and remotely, the data being stored in conventional databases, or in files external to the database, or on the World Wide Web.

7.3 CHALLENGES ARISING FROM BIOLOGICAL DATA

The first problem was the many differing formats in which biological data are found, ranging from the regular structured data of GenBank or PDB files, to data of complex structure such as pathway data, or unstructured data such as literature abstracts. This is a problem for a DBMS which only supports a simple relational table model of data. However, many DBMSs have wider capabilities such as the following.

- Object-oriented DBMSs do have facilities for defining objects having a complex structure, together with the operators for storing and retrieving those objects. Hence, sequence or pathway objects could, for example, be defined and manipulated with such a system. Other DBMS products, known as *object-relational* DBMSs, have the ability to define such objects while retaining tables as the primary representation of data in the database.
- Some DBMS products have special facilities for holding and managing unstructured text data such as are found in biomedical literature abstracts.
- *Semi-structured database* techniques have been developed to handle data which, while not wholly unstructured, do not have the regular structure making them suitable to hold in table form. Currently, XML techniques are often used to handle such data.

The second problem was that there is often no single biological object to which data refer: individual users of the data will have their own perspectives. This is equally the case in considering data concerned with biological functions and mechanisms.

- This remains a problem with many DBMS products. The relational model provides a fixed table representation of the data although view tables can go some way to providing alternative views of the data. More flexible views can be obtained by holding the data in a semi-structured form, with table views being generated as required.

The third problem was the many incompatible formats and nomenclatures for biological data. The key to overcoming this is agreement on standards for representing biological data and their semantics. There are a number of initiatives in this area. For example:

- The *GO* (*Gene Ontology*) consortium has developed a controlled vocabulary for terms describing gene products from three perspectives: molecular function, biological process and cellular component. The ontology explicitly defines the relationship between more specific and more general terms in graph structures for each of these three perspectives. The ontology may be obtained in XML format, and also exists as a relational database which may be queried from a Web-based interface.

- A very large number of XML-based definitions and other resources have been published. These range from XML representations of specific data resources such as GO, InterPro and PIR, to proposals for XML representations of sequence data (BIOML, BSML), BLAST output (BlastXML) and expression data (MAGE-ML), as well as more general XML definitions for ontology specification (DAML+OIL). One example of XML being used in a practical way is DAS (Distributed Annotation System) which uses an XML representation of sequence annotation information to enable integration of annotations retrieved from multiple sites.
- In the object database world, the OMG LSR provides a focus for similarly defining standards and interfaces supporting a common format for biological data.

The final problem was the fact that data about any given biological object are likely to be held at many different sites, each with a different approach to coping with the previous problems.

- Techniques for retrieving data from remote sites are well established. Once again, the problems are those of being able to treat those data as a single integrated resource, and the key is agreement on standards for representing data and their semantics.

7.4 CONCLUSIONS

The data management techniques outlined in this chapter will undoubtedly be developed further to enable distributed biological data to be used as an integrated resource. Many of the most difficult outstanding problems arise from the semantic integration of such data rather than any technical problems in its storage and access as such.

Currently the World Wide Web enables distributed data to be accessed and displayed in a browser, but a program cannot extract the meaning of the data. The term *semantic web* is used to describe what data management techniques will increasingly support: ways of enabling distributed data to be accessed in a form whereby the semantics of the data are explicit and may be extracted and exploited by computer programs. In addition, the computer programs themselves will be distributed resources on the Web, rather than being written by individual scientists to run on local machines. This is the aim of *grid* computing: turning the Web into integrated data and computational resources which a scientist can plug into in the same way that the electricity grid can be plugged into when electrical power is required.

8

Elements of Bioinformatics

8.1 CONCEPTUAL MODELS OF PROTEIN STRUCTURE

8.1.1 Structural Types and Conceptual Models

Conceptual models can add enormously to our understanding of protein sequence, structure and function. It is the purpose of this topic to introduce some key conceptual ideas that are essential to understand protein bioinformatics. It is useful to distinguish three different protein structural types: fibrous proteins (e.g. collagen), *globular proteins*, which tend to exist in aqueous solvents like the cytosol and extracellular fluids, and *integral membrane proteins*, which exist within the lipid environment of biological membranes. In this text we will be concerned mainly with globular proteins and integral membrane proteins.

8.1.2 Globular Proteins

Proteins with a wide variety of functions fall within this broad class, including enzymes, antibodies, and a variety of molecules associated with signaling and transport. Overall they tend to have globular (spheroidal) shapes. Structure formation in globular proteins is dominated by the *hydrophobic effect*. Nonpolar chemical compounds with few charged atoms are *hydrophobic*: they do not dissolve easily in water. In contrast, polar compounds with charged atoms form electrostatic and hydrogen bonding interactions with water molecules and therefore tend to dissolve easily. Some of the amino acid side-chains are hydrophobic (valine, leucine, etc.), while other amino acids are hydrophilic (aspartic acid, lysine, etc.). When a protein folds, it is able to minimize its free energy by placing hydrophilic amino acids on the surface of the globule, in contact with the aqueous solvent, and hydrophobic amino acids in the central core of the globule, away from the solvent. It is thought that this effect is the strongest force driving the linear polymer of amino acid residues to fold into globular shape in water. Although most core side-chains are hydrophobic, there are sometimes some hydrophilic side-chains in the core, very often making hydrogen bonds or ionic interactions with other core side-chains, and equally, there is often a small proportion of hydrophobic side-chains on the protein surface. An important feature of protein cores is that they are efficiently packed, so that most of the available space is filled with atoms.

8.1.3 Secondary Structure

Globular proteins usually contain elements of regular *secondary structure*. The best-known example is the *α-helix*, where four or more consecutive amino acid residues in the polymer adopt the same conformation, resulting in the adoption of a regular helical shape by the polypeptide backbone. This helix is stabilized (held together) by hydrogen bonds between the main chain C=O group of each amino acid and the H-N group of the amino acid four residues further along the helix. This forms a helix with 3.6 amino acid residues per helical turn. Although helices with slightly different hydrogen bonding patterns (3_{10} and π helices) do occur in protein structures, α helices are by far the most common. The illustration shows that the amino acid side-chains point away from the helical axis, forming a surface for the helix. The notation H is used to indicate that a particular residue is a member of an α-helix.

The other common type of regular secondary structure is the β-strand. This secondary structure element is formed by consecutive amino acids in their most extended conformation, and for this reason the letter E (Extended) is used to indicate that a residue adopts this structure. In this case, hydrogen bonds between main chain C=O and N–H groups form not to residues in the same strand, but to residues in strands formed by other parts of the polypeptide. These hydrogen bonds mean that single beta strands do not exist in isolation, but are always spatially adjacent to at least one other strand. The twisted, pleated structure formed by consecutive, spatially adjacent, hydrogen-bonded strands is known as β-pleated sheet. If such a sheet is curved round so that the stands that would have

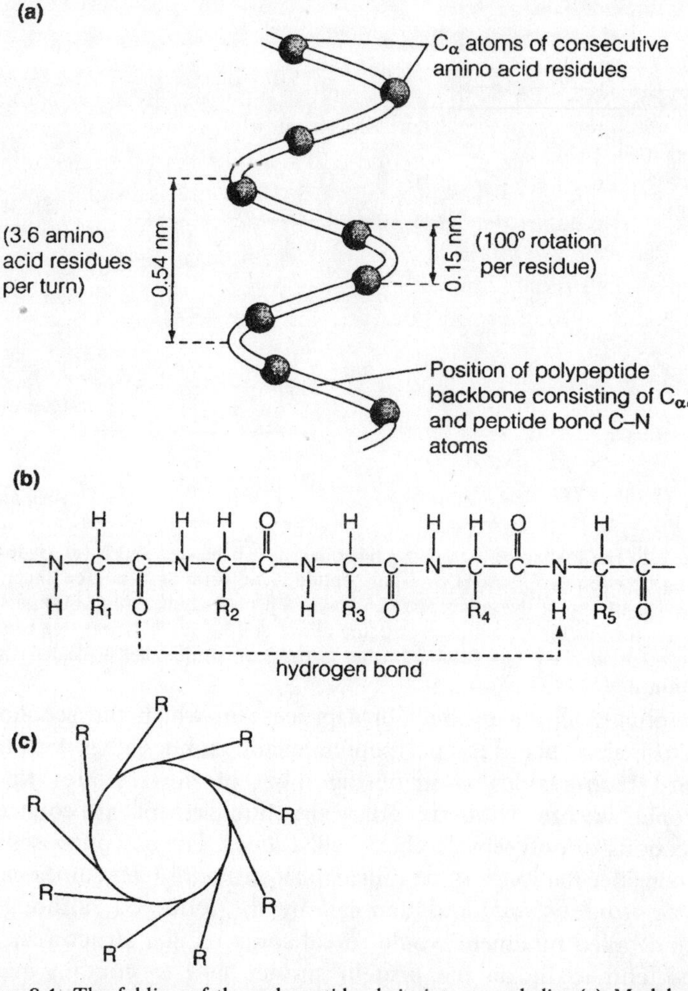

Figure 8.1. The folding of the polypeptide chain into an α helix. (a) Model of an α helix with only the C_α atoms along the backbone shown; (b) in the α helix the CO group of residue n is hydrogen bonded to the NH group on residue (N+4); (c) cross-sectional view of an α helix showing the positions of the side chains of the amino acids on the outside of the helix.

been on the edges of the sheet are spatially adjacent and hydrogen bonded, the structure is known as a β barrel.

8.1.4 Tertiary Structure

The *tertiary structure* or *fold* of the protein is the position of every protein atom in three-dimensional space. This can be considered

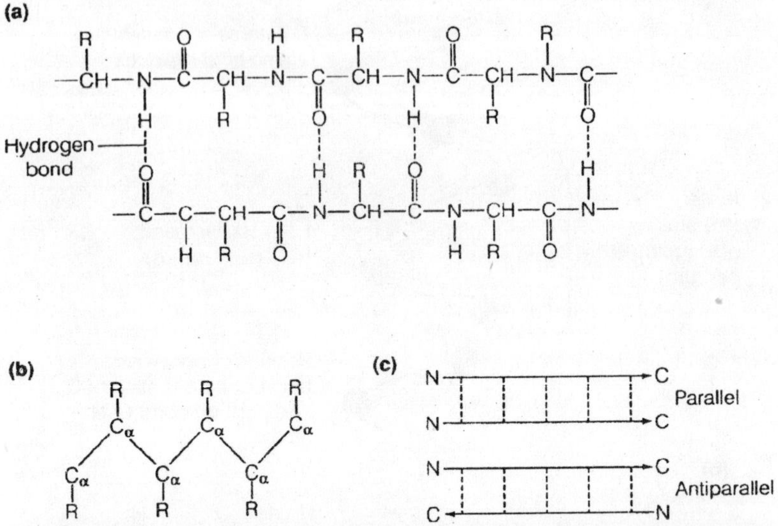

Figure 8.2. The folding of the polypeptide chain in a β-pleated sheet. (a) Hydrogen bonding between two sections of a polypeptide chain forming β-pleated sheet; (b) a side-view of one of the polypeptide chains in a β-pleated sheet showing the side-chain attached to the C_α atoms protruding above and below the sheet; (c) because the polypeptide chain has polarity, either parallel or antiparallel β-pleated sheets can form.

conceptually as the product of a process in which the secondary structure elements of the polypeptide chain pack together with their inward facing sides contributing most of the residues to the hydrophobic core. These secondary structure elements are connected by sections of polypeptide chain called *loops*. For our purposes, we can consider the loops to be of irregular structure (sometimes called *coil* or *random coil*, and denoted by the letter *C*), although a more detailed treatment would reveal some regular structures. The loops tend to lie on the protein surface and to contain mainly hydrophilic residues.

8.1.5 Quarternary Structure

Many proteins exist as multimeric molecules formed as several polypeptide chains bind together to form a complex. This complex is known as the quaternary structure of the protein, and it is the biologically functional unit. Sometimes complexes have a single biochemical function, but there are many examples of multifunctional complexes, for instance enzymes that catalyze consecutive steps in a metabolic pathway.

8.1.6 Integral Membrane Proteins

Integral membrane proteins are key elements of biological signaling and trans-membrane transport systems. Examples are the G protein-coupled receptors (involved in signaling), and channel proteins responsible for rapid transport of ions across membranes. The fact that significant parts of these proteins exist in the lipid environment of biological membranes, in contrast to globular proteins that exist in aqueous solvents, means that different structural principles apply to them. Typically, integral membrane proteins have one or more segments that actually cross the membrane. Amino acids in these segments tend to be of hydrophobic nature, compatible with the lipid structure of the membrane, while portions of the protein that exist in the more aqueous environment on either side of the membrane tend to be more polar. Very often the membrane-spanning segments adopt an a-helical structure, as in G protein-coupled receptors and many ion channels, but there are examples, for instance the bacterial porins, where the membrane-spanning segments are β strands. The number of membrane-spanning sections within a typical membrane protein varies from one (often a membrane 'anchor' for the protein), to more than 10. The well-known G protein-coupled receptors, like rhodopsin, contain seven membrane-spanning segments.

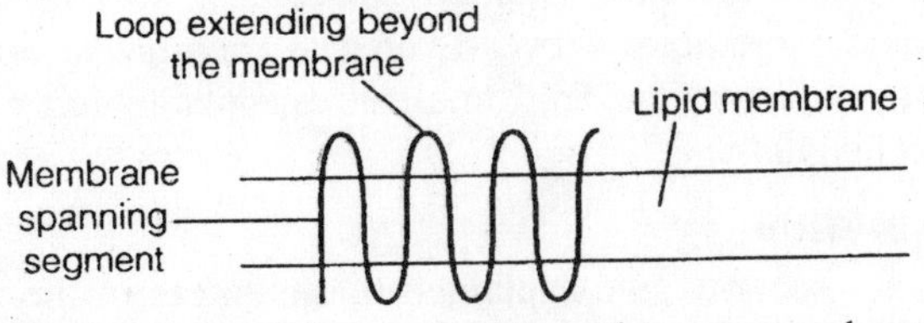

Figure 8.3. An integral membrane domain with seven membrane-spanning segments.

8.1.7 Domains

From the knowledge of protein sequence and structure we have now, it is clear that many proteins have a modular architecture. Nature creates proteins with complex functions by combination of quasi-independent modular units or *domains*, typically with much simpler functions. For instance there are many eukaryotic proteins that contain the homeo domain whose function is to bind to DNA, but these proteins also contain other domains that are responsible for other aspects of their overall function. The precise definition of a protein domain depends on your point of view. Some would define

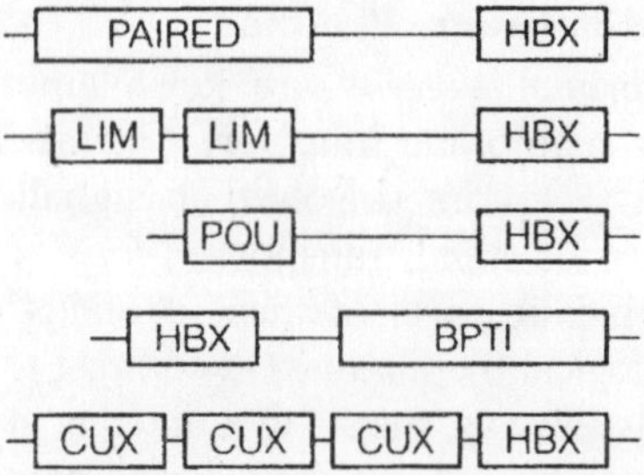

Figure 8.4. Multidomain proteins containing a homeo domain.

it as a protein sub-sequence or substructure with a recognized function, as above. Others view domains as protein sub-sequences that are able to fold independently of the rest of the protein, and others regard domains as geometrically distinct parts of the protein structure. Perhaps the most useful definition of a domain, however, is that it is a protein unit observed to occur in many otherwise unrelated proteins.

Whichever your preferred definition of a protein domain, it is very important to bear in mind the modular architecture of proteins in many bioinformatics analyses. It is also important to appreciate that proteins can have domains of mixed types, for instance integral membrane domains are often found in combination with globular domains. This is the case for many receptors, which are anchored to a biological membrane by an integral membrane domain, and have extracellular globular domains responsible for recognizing biochemical signals.

8.1.8 Evolution

In earlier sections, we explained that changes in the amino acid sequences of proteins occurring during evolution are accepted if they have a neutral effect on protein structure and function (Le. they have little effect on the stability of the protein and its ability to perform its normal function), or if they have a positive effect (i.e. they enhance stability or functional efficacy). When a change has a negative effect it is likely to be rejected by natural selection, except in the cases of the changes occurring when a duplicated gene is in the process of evolving a new function.

When we observe the evolution of protein structures we see that evolutionary change is very slow for residues in the structural core, and significantly faster for residues on the protein surface. This can be understood in light of the conceptual models introduced above. Within the structural core the introduction of changes can affect the tight atomic packing, resulting in reduction of protein stability, and is therefore less likely. On the other hand, residues on the protein surface are not subject to this constraint (essentially they need only to be hydrophilic) and are therefore much more susceptible to evolutionary change. Because secondary structure elements tend to contribute most of the core residues, while loop residues lie mainly on the surface, it is often observed that loops evolve much more quickly than secondary structure elements. A related effect is observed in integral membrane proteins, where the membrane-spanning segments (most often helical) evolve more slowly than the loops that connect them.

8.2 RELATIONSHIP OF PROTEIN THREE-DIMENSIONAL STRUCTURE TO PROTEIN FUNCTION

8.2.1 Structure and Function

The ability of proteins to perform biological functions depends on the formation of a three-dimensional structure (fold) that is stable in the normal environment of the protein. For example, enzymes often catalyze reactions using an *active site*. Typically, this is a cavity in the three-dimensional structure of the enzyme that is accessible to the reactants from the protein surface. Active sites are multifunctional. They contain the key catalytic machinery of the protein, which is typically one or more residues that are actively involved in the chemical reaction catalyzed, and stabilize transition states. They exclude solvent from the reaction, and their shape and physicochemical properties are such that they bind the intended reactants much more strongly than any alternatives, thereby creating catalysis specific to certain molecules or molecular classes. All this depends on the active site adopting a specific three-dimensional shape, and ultimately on the fold of the peptide chain itself.

Not all proteins are enzymes, but most rely on their three-dimensional structures in order to perform biological functions. Molecular recognition without catalysis is a function of many proteins. Transport proteins need to recognize the molecules they carry. Protein-protein recognition is important in the recognition of foreign proteins by antibodies, the interactions between components of signaling pathways, and the formation of multifunctional complexes. Similarly, the recognition of other macromolecules by proteins is important in the regulation of gene expression (by DNA binding

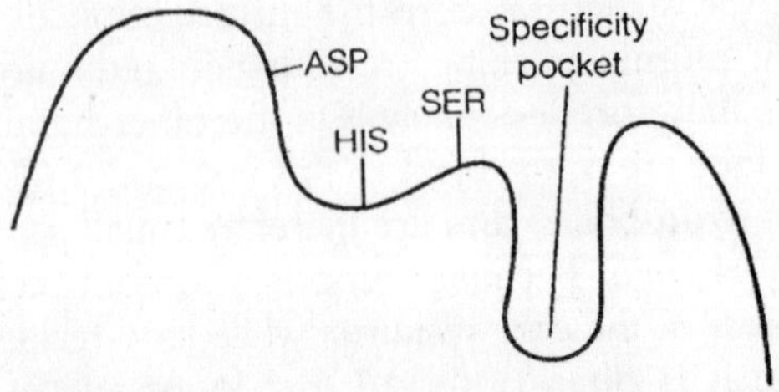

Figure 8.5. A schematic view of a serine protease active site where the line represents the protein surface.

proteins), and the formation of mixed protein-RNA complexes like the ribosome. Finally many receptor proteins need to specifically recognize a molecular signal (for example, the recognition of steroid hormones by receptors in the cell nucleus).

For molecular recognition to take place, it is necessary that the molecules concerned are able to bind together in an energetically favourable conformation. This depends on *complementarity*. They must be able to form complementary shapes, so that they fit snugly together, and where their surface atoms are in contact there must be complementarity of physicochemical properties. This means that negatively charged areas of one interacting partner must contact positively charged areas of the other, and hydrophobic areas must be in contact with other hydrophobic areas. All this depends on the formation of a stable three-dimensional structure by the protein.

8.3 EVOLUTION OF PROTEIN STRUCTURE AND FUNCTION

8.3.1 Structural and Functional Constraints

The evolution of protein sequences was already discussed, where amino acid substitutions were understood in terms of physicochemical relationships between the different amino acids. Amino acids with similar size and physicochemical properties are likely to make reasonable replacements for each other, and to form accepted substitutions during protein evolution. With structural knowledge, we can be more precise. Fundamentally, amino acid substitutions are accepted in evolution if the change is either neutral or advantageous to protein structural stability and/or function. Although matrices such as PAM250 can tell us about average rates of substitution between amino acids, when the three-dimensional structure is known we are able to understand which substitutions are acceptable at particular positions in the protein sequence.

Some amino acid residues have key roles to play in the stability of particular structures. When a residue plays a key role it is often the case that no other residue can substitute for it while maintaining the stability of the structure, and the residue is conserved throughout the evolutionary family. In a similar way, key functional residues are also often conserved. Many enzymes rely on the chemical properties of certain amino acids to effect catalysis, for instance serine, histidine and aspartic acid residues in the serine protease active site. If these properties are unique to the residue involved, it will tend to be conserved to preserve function.

Table 8.1. Unique structural roles played by some amino acid residues in protein structures.

Residue	*Structural role*
Cysteine	The formation of disulfide bridges (S-S bonds) with other cysteine residues, making covalent links between sequence separated parts of the polypeptide chain. These bonds can make a major contribution to structural stability.
Glycine	The side chain of glycine is very small (a single hydrogen atom). This means that the structure of glycine is more flexible than that of the other amino acids. It can adopt some conformations that are accessible to no other residues, and is often found where the peptide chain goes through a very tight turn.
Proline	Amino acids tend to prefer to adopt *trans* conformations around the planar peptide bond, that is, with the two a carbons located on opposite sides of the bond. *Cis* conformations (with the acarbons on the same side of the bond) occur occasionally, and often involve a proline residue, which is better able to form the Cis conformation than the other amino acids.

8.3.2 Multiple Sequence Alignment

The relationship between multiple sequence alignments and protein structure and function is very important. Many of the features of the alignment can be understood in the light of a known structure for one member of the family.

We already explained that residues on the protein surface evolve more quickly than residues buried in the structural core, because they are not subject to the strong constraints involved in the maintenance of efficient packing of the structural core. We also described a conceptual model of globular protein structure in which secondary structure elements (helices and strands) contribute most residues to the hydrophobic core of the protein and alternate with loops whose residues lie predominantly on the surface. Loops are therefore observed to evolve more quickly than secondary structure elements. This effect is observed in multiple sequence alignments, where it is often clear that the alignment consists of alternating blocks of well-conserved residues corresponding to secondary structure elements and less well-conserved residues from surface loops. This is evident in the alignment, where it is clear that insertions and deletions in the sequences fall in loops (outside the labeled secondary structure elements), and that strong conservation of hydrophobic residues is observed in some secondary structure elements. For example, the second helix contains a group of three residues of hydrophobic character, which are W, V and C in the first (top) sequence.

Multiple alignments made on the basis of sequence alone by most standard software methods are often not perfect. Structural information can be used to make manual improvements. For instance, if it is known that a particular cysteine is involved in a disulfide bridge it should be verified that it is aligned with a cysteine in all sequences in the family. Similarly, if the alignment involves insertions and deletions, it should be borne in mind that these are much more likely in loops than in secondary structure elements.

In the case where there is no known three-dimensional structure, a multiple alignment can serve as a first stage in structure prediction. Conserved cysteines can be predicted to be involved in disulfide bonds (although this is not the only reason for conservation of this residue, which is also often used to bind a metal ligand). Similarly, the alternation between blocks of conserved residues and blocks where the sequences are more variable, which is apparent in many multiple alignments, can be interpreted in terms of probable secondary structure elements alternating with surface loops.

Another structural feature that is often visible in multiple sequence alignments is the amphipathic helix. The amino acid side-chains point away from the helical axis and form a surface for the helix. Very often, a helices are positioned in protein structures so that one side of the helix is part of the hydrophobic structural core and the other side lies on the protein surface. Hydrophobic side-chains therefore

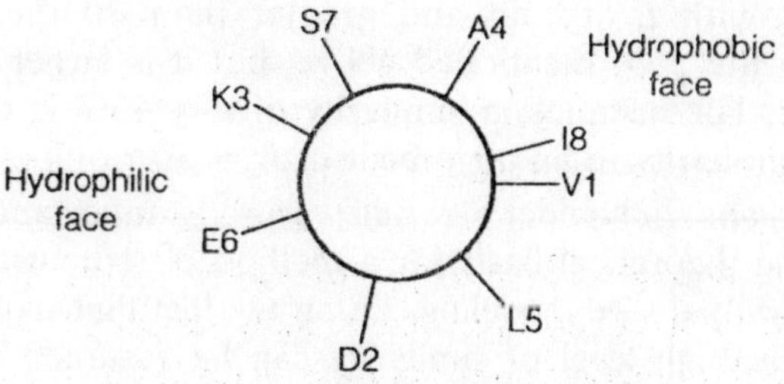

Figure 8.6. An amphipathic α helix.

dominate on one side and hydrophilic ones on the other. Since an a helix has 3.6 residues per turn, an amphipathic helix tends to exhibit an alternation between hydrophilic and hydrophobic residues with a periodicity of three to four residues. This is often visible in multiple sequence alignments as conserved hydrophobic residues occurring alternately every three or four residues in the sequence.

8.3.3 Evolution of the Overall Protein Fold

So far, we have considered the evolution of parts of the protein structure in isolation, now we consider how the overall structure evolves. In 1991, Sande and Schneider published a study of naturally occurring proteins of known structure showing how much similarity in protein sequence is needed to guarantee similarity in structure. They found that the threshold of similarity depended on the lengths of the sequences involved, with more similarity being required for shorter sequences. They showed that if two protein sequences can be aligned to show more than 25% identical residues over an alignment length of more than 80 residues, then they will share the same basic three-dimensional structures. This result only applies to the naturally occurring proteins currently of known structure. There are now examples of artificially engineered proteins with sequence similarity of as much as 50% yet with different structures. Other potential exceptions to Sander and Schneider's rule are proteins like the prion protein that appear to have two distinct possible structures (one normal and the other disease associated). The Sander-Schneider rule was derived from a data base in which the proteins were in their normal biologically active conformations, and these are the conformations to which it applies.

Sander and Schneider's threshold for structural similarity was

$$t\,(L) = 290.15L^{-0.562}$$

where L is the length of the alignment and t is the percentage identity threshold required to guarantee structural similarity. For alignments with L around and greater than 80 the value of t approaches the 25% mentioned above, but it is larger for smaller values of L. For instance, a similarity of t = 43% is required for structural similarity in an alignment of L = 30 residues.

Sander and Schneider's result was very important because it provided the theoretical basis for a method of structure prediction known as comparative modeling. Using the fact that sequences with a sufficiently high level of similarity can be assumed to have the same structure, sequences of known structure can be used in structure predictions for sufficiently similar sequences of unknown structure.

8.3.4 Conservation of Structure

Sequences with sufficient similarity share the same structure, but observations made on known sequences and structures have shown us that divergent evolution of proteins with a common ancestor can operate until their sequence similarity is almost undetectable. The best-known example of this comes from the globin family, including the haemoglobin chains, myoglobin, and plant leghaemoglobin. These sequences adopt the same overall structure, and carry out similar functions (oxygen transport) by the same mechanism, but often exhibit sequence identities that are below 20% of identical residues. This is a level of sequence similarity that is often found between the sequences of completely unrelated proteins. This observation that sequence evolves much faster than structure means that structure is very important to bioinformatics. Many potential evolutionary relationships that were not apparent on the basis of sequence analysis have been discovered when three-dimensional structures have become available.

8.3.5 Evolution of Function

We have commented in previous sections that protein function evolves, so that proteins that are clearly homologous (i.e. have closely related sequences and/or structures) sometimes have different functions, and we have identified gene duplication as a possible mechanism for the development of new functions. When this happens, residues that have a key role in the performance of one function that are conserved throughout the members of the family performing that function are often not the same or even conserved in members of the family which perform a different function. Lysozyme and α-lactalbumin are examples of homologous proteins that perform very different functions (the former is an enzyme and latter a regulatory protein). The multiple alignment the key lysozyme functional residues are not conserved in α-lactalbumin, while key structural residues are (in particular the disulfide bond-forming cysteine residues).

Gene duplication is not the only way in which protein function evolves. Multidomain proteins are proteins that consist of more than one domain. Often the different domains have different functions that are combined to produce another, much more complex function. For example, some enzymes have domains associated with regulation of their activity. It seems that nature evolves many new proteins by swapping and recombination of modular units (domains) and that this is a major route for the evolution of new and more complex protein functions.

8.4 OBTAINING, VIEWING AND ANALYZING STRUCTURAL DATA

8.4.1 Software and WWW Sites

It would be impossible to produce a comprehensive list of available software and World Wide Web (WWW) sites for the display and analysis of protein and other macromolecular structures. All we provide here are pointers to the most commonly used and freely available utilities. The software and WWW sites we discuss are all linked from the text WWW site. It should be noted that this book is not intended to serve as a software manual for commonly used programs; that function is much better provided by the documentation that is distributed with the software itself. Such documentation will be much more extensive than the space available here would allow, and should remain up to date with changes to programs as they are made.

8.4.2 Obtaining Data

We have already discussed the structural database and the experimental methods that produce the data contained within it. Like the sequence databases discussed in earlier chapters, the structural database can be searched by many of the standard search tools, including SRS and tools available at the National Center for Biotechnology Information (NCBI) WWW site. There are also specialized search engines maintained by the macromolecular structural database group at the European Bioinformatics Institute (EBI; http://msd.ebLac.uk),and the Research Collaboratory for Structural Bioinformatics (RSCB; http://www.rcsb.org/pdb). Both these tools allow easy downloading of structural data in the standard Protein

Data Bank (PDB) format, which is input to much of the software discussed below.

8.4.3 Visualization of Structures

The program *RasMol,* written by Roger Sayle, is perhaps the bestknown viewer for macromolecular structures. Given a structure in the standard structural database format, this software displays a three-dimensional image of the structure. The image can be rotated by use of the mouse to produce different views, and displayed in various formats, including *wireframe* (bonds displayed as lines with atoms implied at junctions and ends), *space filling* (each atom represented by sphere of proportionate size), and *ball* and *stick* (atoms represented by small spheres and chemical bonds by sticks). There are also special *cartoon* formats that give clear displays of secondary structure elements. The user can choose between various colour schemes and even use customized colours. Finally, there are flexible ways of selecting parts of structures to enable highlighting with a different display format. For instance, key residues or substructures can be highlighted by variation in the display mode. Small protein structure displayed in RasMol in two different display formats, the first showing all atoms, and the second abstracted to show only secondary structure elements, with helices as helical ribbons and stands as extended ribbons. A more complex example of an enzyme with its inhibitor displayed in different formats along with the RasMol commands that produced the display.

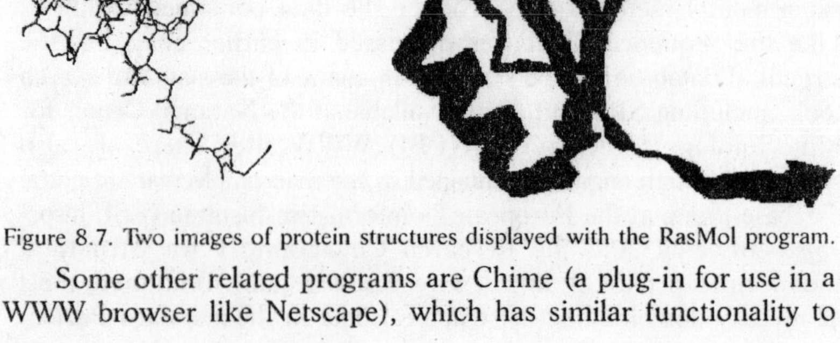

Figure 8.7. Two images of protein structures displayed with the RasMol program.

Some other related programs are Chime (a plug-in for use in a WWW browser like Netscape), which has similar functionality to

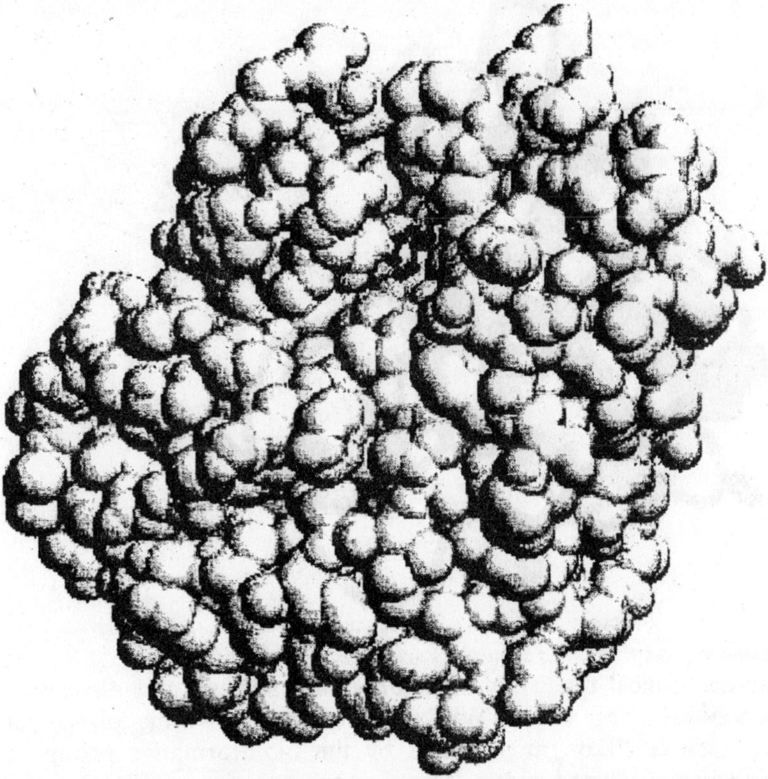

Figure 8.8. The structure of the serine protease enzyme trypsin in 'space-filling' format with an inhibitor (benzamidine) shown in black, bound in the main specificity pocket of the enzyme adjacent to the active site.

RasMol, and Cn3d, which is also able to link the structure display directly to a multiple sequence alignment. The visualization of protein structures in three dimensions can be very difficult, and for this reason there are several tools that produce schematic or summary displays of protein folds. One example of this is the RasMol cartoons format, and similar more sophisticated displays can be produced with Molscript and Ribbons. Finally, TOPS is a utility that produces two-dimensional summaries of three-dimensional protein folds.

8.4.4 General Structural Analysis

Protein structures offer almost unlimited possibility for analysis programs, including the automatic annotation of secondary structure elements, the analysis of residue solvent accessibility to find surface and core residues, the creation protein surfaces, checking the

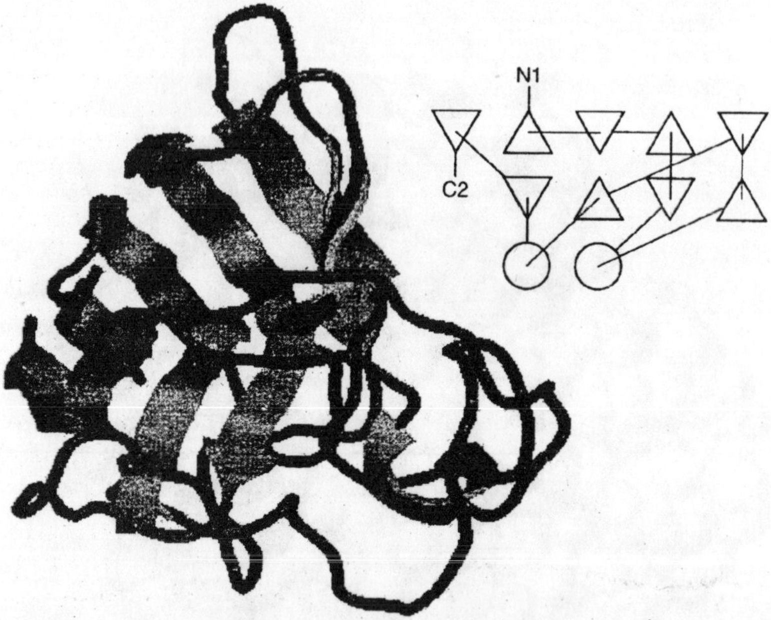

Figure 8.9. A TOPS protein-folding topology diagram (right) for a superoxide dismutase and the full structure in RasMol cartoons format (left).

stereochemical quality of structures, and analysis and display of functional sites. One WWW resource that is worthy of special mention is PDBsum provided by the bioinformatics group at University College London. This resource contains a well-presented summary of structural data and analyses for every entry in the structural database, accessed by the accession code for that database. The information presented includes secondary structure, disulfide bonds, ligand binding sites, active sites, key residues, plots of intermolecular interactions, folding topologies, EC numbers for enzymes, and much more.

8.4.5 Analysis of Functional Sites

Probably the most important parts of any protein structure are its functional sites, that is, active sites of enzymes, ligand binding sites, metal ion binding sites, sites where other proteins or DNA might bind, etc. Location of such sites in the database of known structures is surprisingly difficult. Some sites are annotated by those who determined the structures, and others can be located because they contain bound ligands, and most of these are part of PDBSum. Nevertheless, many potentially interesting sites in known protein structures cannot be found in this way, and there is a need for methods that can locate such sites as part of the functional analysis of new protein structures.

To analyze a new structure it is necessary to first locate potential functional sites and then consider what the structure of these sites can reveal about, possible functions. It has been observed that enzyme active sites are often located in large cavities near the protein surface (more often than not the active site is the largest such cavity). The SURFNET software is able to locate such sites. To make a prediction about the actual function of any potential site on a protein surface is a problem currently at the forefront of bioinformatics research. In common with much of bioinformatics, most current approaches are knowledge based, and rely heavily on databases of information about active sites of known function. This is based on the idea that active sites that are in some sense similar are likely to have similar functions.

There are several databases, computer programs and resources that implement similarity searches for functional sites in proteins. Such similarity searches are both geometrical and chemical: they search for sites where the relative geometrical positions and chemical type of atoms, residues or chemical groups are similar. PROCAT is a manually curated database of three-dimensional atomic active site templates, which define particular enzymatic activities. These have been extracted from the database of known structures, and a new protein structure can be scanned against the database to search for the presence of a known catalytic template. ASSAM and SPASM allow the user to define interesting residues in a protein structure and search similar sets of residues in the same relative spatial positions in other known structures. In a similar vein, Relibase is a database of protein-ligand complexes with sophisticated geometrical search facilities. All these tools are linked from the text WWW site, and form a formidable armory for functional analysis of new protein structures.

8.5 STRUCTURAL ALIGNMENT

It should be obvious that the alignment of protein or nucleic acid sequences is a very important part of bioinformatics. An alignment defines a relationship between the elements of one sequence and those of another, and the degree of similarity of the sequences reflected in the alignment is the basic data on which we base judgments, by statistical or other means, of the likely biological meaning of the relationship. Proteins are likely to have similar structures and biological functions if their sequences can be aligned to show significant similarity.

Sequence alignment is relatively easy when the sequences are closely related, involving a high proportion of identical monomers, and few insertions and deletions. However, as we have commented earlier in this section, the sequences of evolutionarily and structurally related proteins can be very different. Divergent evolution can lead to sequences whose similarity is almost unrecognizable. In these cases, naive sequence alignment based on information only from within the sequences and standard substitution matrices can be very difficult. It can be almost impossible to find an alignment that is even approximately correct in biological or evolutionary terms.

Knowledge of protein structure provides a means by which highly divergent sequences can be aligned, and this process is known as *structural alignment*. This can be described conceptually as a process in which two similar structures are superimposed in three dimensions, so that peptide backbones of structurally equivalent residues lie close together in space. This superposition is then used to define a sequence alignment in which the aligned residues are structurally equivalent. Because structure is more strongly conserved during evolution than sequence, the structural alignment is much more likely to be correct in terms of biological function and evolution.

Two alignments of a pair of distantly related globin sequences are shown—the first produced by sequence only means and the second a structural alignment. The alignments are different, and although the sequence only alignment shows a higher percentage of identical residues it is certainly not as good as the structural alignment. One very obvious advantage of the structural alignment is that the

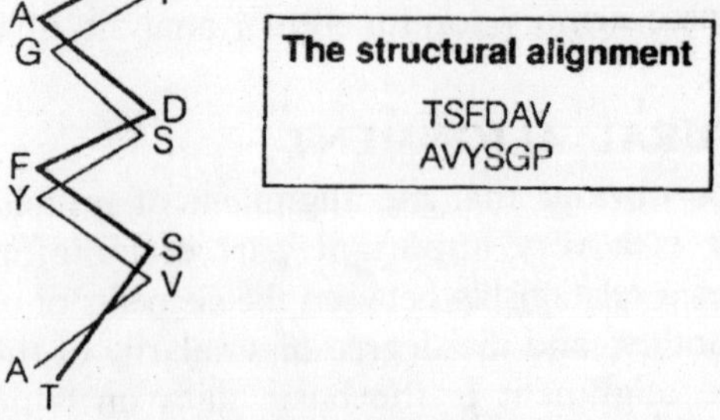

Figure 8.10. A conceptual view of structural alignment.

histidine residues that coordinate the heme iron in both globins are aligned (equivalenced), whereas they are not in the sequence only alignment.

8.5.1 Software

There is a large amount of software available to perform structurally based alignments, and some of this is linked from the text World Wide Web (WWW) site. A good example is DALI WWW (http://www.ebi.ac.uk/dali). The detailed methods used by the various computer programs are different, and almost all of the methods do not conform exactly to the simple conceptual description of structural alignment given above. Some methods make use of sequence as well as structure information, and almost all make use of information about secondary structure. There is no real consensus about which methods are the best, but any alignment of distantly related protein sequences will be better if created using structural information.

8.5.2 Structural Similarity

Most structural alignment methods produce measures that can be used to assess the level of similarity of the aligned structures. A variety of such measures are used, many of which are specific to particular structural alignment methods. Nevertheless most methods report a measure known as *root mean square deviation* (RMSD). When two structures have been aligned, they can be optimally superimposed so the aligned residues lie as close as possible to each other in three-dimensional space. With the structures superimposed, distances can be measured between the a carbon atoms of the aligned residues, and the root mean square deviation is defined in terms of these distances by

$$RMSD = \sqrt{\frac{1}{N}\sum_i d_i^2}$$

where d_i is the distance between the ith pair of superimposed (aligned) a carbon atoms, and N is the number of such superimposed atoms (i.e. the number of aligned residues). This expression represents the square root of the mean square distance between aligned residues. If the structures were identical then the RMSD would be zero, because all residue α carbon atoms would lie directly on top of one another (zero distance apart), after optimal superposition. For less similar structures, the RMSD becomes larger. Generally RMSD values between 0.0 and 1.5 Å represent very similar structures, with higher values indicating progressively increasing structural dissimilarity. Clearly a small RMSD computed over a large number of residues (N) is more significant than a small RMSD computed over a small number of residues.

8.5.3 Structural Similarity Searches

Just as we are often interested in searching sequence databases for sequences similar to a query sequence, it is sometimes desirable to search the structural database for structures that are similar to a query structure. A number of WWW sites provide this type of search, including DALI, SSAP, TOPS, VAST and the Research Collaboratory for Structural Bioinformatics (RCSB) site (see the text WWW site for links). As with sequence searches, these search engines return a list of similar structures ranked according to some measure of similarity, such as RMSD.

8.6 CLASSIFICATION OF PROTEINS OF KNOWN THREE-DIMENSIONAL STRUCTURE: CATH AND SCOP

8.6.1 Why Classify Protein Structures?

To classify protein structures means to group them so that each group contains similar structures. As stated already that naturally occurring protein sequences can be grouped into evolutionary families. It has already been discussed that protein structure is much more strongly conserved by evolution than protein sequence. To classify proteins by structural criteria is therefore the most powerful way to assigning them to families, and to reveal distant evolutionary relationships.

8.6.2 Structural Class

At the top level, each classification places protein structures into broad structural classes. CATH uses four classes reflecting secondary structure content: all α, all β, α and β ($\alpha\beta$), and few secondary structures. SCOP on the other hand has more divisions, separating out more small groups, like membrane and coiled-coil proteins, and also splitting the α and β class into two, $\alpha + \beta$ and α/β, depending on whether the α helices and beta strands are segregated in the fold ($\alpha + \beta$) or mixed up and tending to alternate (α/β). The next level in CATH further separates the structural classes into broad architectures reflecting the overall shape of the protein, and has no equivalent in SCOP.

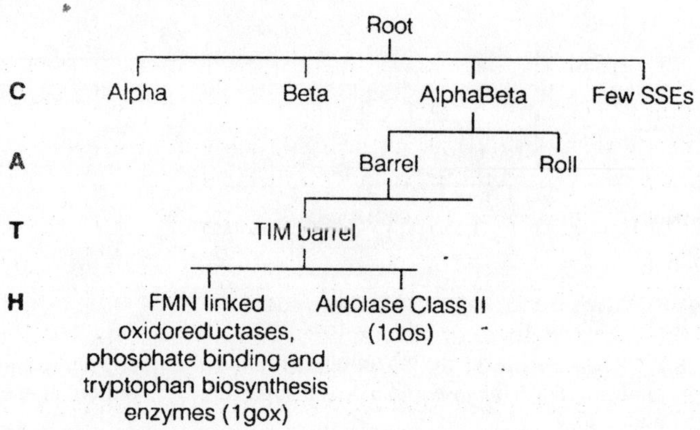

Figure 8.11. The structure of the CATH classification illustrating the positions of the glycolate oxidase structure with PDB code 1gox and the class II aldolase structure 1dos, which share the TIM barrel fold.

8.6.3 Fold or Topology

The CATH T (topology) and SCOP fold levels separate proteins into groups that have the same overall fold. This means that the proteins have the same core secondary structure elements (SSEs), connected and arranged in space in more or less the same way. The examples given are part of the TIM barrel-fold level. Within this level, all proteins have the well-known TIM barrel structure: a parallel eight-stranded beta barrel surrounded by α helices. When proteins have the same fold or topology, they can usually be structurally aligned to give large sections of superimposed backbone structure with low root mean square deviation (RMSD) values.

8.6.4 Homologs and Analogs

It is thought that many proteins with the same fold have emerged by divergent evolution from a common ancestor, but it is equally possible that they have no common ancestor and adopt the same fold simply because that fold is favourable from a physicochemical point of view. In the former case, the proteins are called *homologs* and are said to be *homologous,* in the latter they are called *analogs,* and are said to be *analogous.* It is almost impossible to know whether two weakly similar proteins have derived from a common-ancestor or not. However, CATH and SCOP divide proteins with the same fold into homologous super-families. Within such a super-family, there is good evidence that the proteins are homologous. This evidence might be striking similarity in sequence, structure or function When proteins with the same fold are in different super-families, it

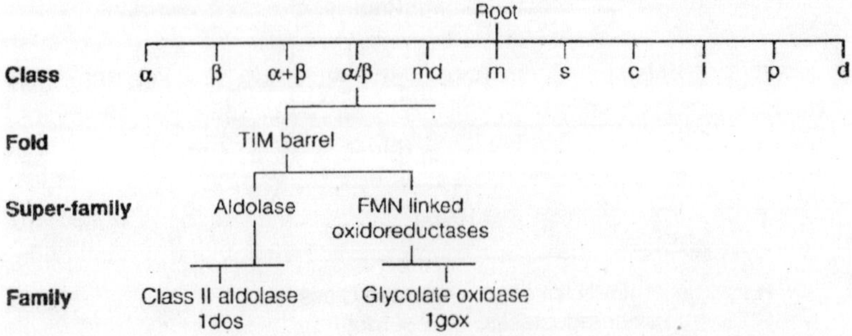

Figure 8.12. The strucure of the SCOP classification illustrating the positions of the FMN linked oxidoreductase and aldolase super-families, and the classification of 1 gox and 1 dos.

means that evidence for homology beyond the shared fold is weak, or nonexistent.

As told already that although the two classifications agree on the fold classification of the proteins in question (TIM barrel), they differ slightly at the homologous super-family level. Both assign the glycolate oxidase and aldolase examples to different homologous super-families, but CATH has a separate super-family for the class II aldolases, which are gathered into a general aldolase super-family in SCOP (containing class I aldolases in addition). This reflects a difference of expert opinion on whether there is sufficient evidence that the two classes of aldolase should be viewed as homologous. It should also be noted that the authors of these classifications do occasionally change their minds about the classification of particular groups of proteins at this level, and this is often reflected in changes made in newer versions of the database.

8.6.5 Super-folds

Only a relatively small number of the different folds in CATH or SCOP contain more than two homologous families. Such folds appear to have arisen more than once in evolution, and are called *super-folds*. Example super-folds are the TIN barrel and the immunoglobulin fold. Characteristics of the super-folds are that they tend to exhibit approximate symmetries, and are characterized by repeated super-secondary structures. These suggest the possible physicochemical favourability of these folds.

8.6.6 Families

Proteins within homologous super-families are often quite distantly related. Both classifications divide each super-family into a number of smaller families within which the relationships between the proteins is much stronger, and often apparent at the level of sequence similarity.

8.7 PROTEIN STRUCTURE PREDICTION

8.7.1 Why Predict Structure?

In the previous topics in this section, we have discussed protein three-dimensional structures and how these enable proteins to carry out their functions. An understanding of structure leads to an understanding of function and mechanism of action. Structural knowledge is therefore vital if we are to proceed to a complete understanding of life at the molecular level. Unfortunately, structural knowledge is still rather limited, because the experimental process of structure determination is slow for most proteins and not possible for many. The database of known structures currently contains more than 18,000 protein structures, but the databases of sequences contain hundreds of thousands of sequences. This gap between sequence and structure knowledge is often termed the *sequence structure gap*; from a practical point of view, it is the main factor motivating the need for predictions of protein structure. To be added to this is the fact that many pharmaceutical drugs act selective binding to target proteins, and knowledge of protein structures can aid the process of rational structure-based drug design.

As well as having enormous practical significance, structure prediction is a fascinating scientific problem that has interested scientists since before the first protein structures were determined. In fact, Pauling predicted the structure of the α-helix before it was observed experimentally. Much of the fascination of the problem stems from the work of two scientists, Anfinsen and Levinthal. By a set of elegant experiments on RNaseA, Anfinsen was able to prove that proteins can fold to their native structures spontaneously, without the intervention of any other agent. The protein fold is therefore somehow coded in the amino acid sequence. Levinthal, on the other hand, pointed out that even relatively small proteins have an astronomically large number of possible structures, and that the process of finding the correct one cannot possibly proceed by a random search of the possibilities because this would simply take too long. To discover how the protein sequence codes for a three-dimensional structure, and to understand how this structure folds has therefore been a problem of much scientific interest for many years. Despite this, the problem is still far from solved, and even though there has been progress, we still do not know in general terms how structure is encoded in sequence.

8.7.2 What is Structure Prediction?

In the most general form, structure prediction means to make a prediction of the relative position of every protein atom in three-dimensional space using only information from the protein sequence. However, not all prediction methods are so general, and some predict limited aspects of structure, for instance secondary structure, without proceeding to predictions of actual atomic positions. In the Topics that follow, we will discuss protein structure prediction methods in four categories, the use of comparative modeling, fold recognition, secondary structure prediction and *ab initio* prediction. We will also discuss the prediction of membrane spanning segments and integral membrane protein topology.

It is useful to categorize prediction methods further according to their theoretical basis, either *ab initio* or *knowledge-based*. Like all physicochemical systems, proteins are believed fold to attain a state of minimum thermodynamic free energy. In our terminology, only methods that attempt to calculate and minimize this free energy, or a suitable approximation, are *ab initio* methods. They proceed from fundamental physical principles, using the accepted theories of quantum mechanics and statistical thermodynamics, to predict protein structures. This is very difficult. Proteins, when modeled with enough solvent molecules to be realistic, are enormous systems with many thousands of atoms. With current technology, detailed calculations of exact free energies are just not possible, and are not likely to be for many years to come. The task therefore is to find a suitable approximation to the free energy that captures the essentials of the folding problem. Such an approximation has not yet been found.

In contrast to *ab initio* methods, knowledge-based methods attempt to predict protein structure using information taken from the database of known structures. The simplest example of this is to predict that a sequence, that is similar to a sequence of known structure, will adopt that same basic structure. This was shown by Sander and Schneider, but there are many other ways of using information from the database of known structures in predictions. The methods of comparative modeling and fold recognition are knowledge-based methods. We also view secondary structure prediction as a knowledge-based method. This is because most secondary structure prediction methods have been trained on data from known structures. However, it is sometimes considered as an *ab initio* method because it is applicable even to sequences that form tertiary structures that have not yet beet observed in any structural relative, however distant. There is no doubt that knowledge-based methods are currently the most accurate and practically useful protein structure prediction methods.

In the past few years, protein structure prediction methods have been subjected to .some rigorous blind testing associated with the CASP (*Critical Assessment of Structure Prediction*) competitions. The results of these competitions have added enormously to our understanding of the accuracy of the various methods. Details of the CASP competitions are linked from the text WWW site at http://predictioncenter.llnl.gov/.

8.8 STRUCTURE PREDICTION BY COMPARATIVE MODELING

8.8.1 Theoretical Basis

As already discussed that two proteins share the same structure if their sequences can be aligned to show at least 25% identity in an alignment of 80 or more residues. This is the basis of the method of *comparative modeling* (also sometimes known as *homology modeling*). If a sequence of unknown structure (usually called the *target sequence*) can be aligned with one or more sequences of known structure, so as to satisfy this condition, then the known structures can be used to predict the structure adopted by the target sequence. The proteins of known structure used are usually referred to as *template structures*.

8.8.2 Ingredients

All that is necessary for comparative modeling is an alignment between the target sequence and the sequences of the template structures. The first step is to locate possible template structures, and this can be done using the standard sequence similarity search methods already discussed. In this case it is only necessary to search the database of sequences whose structures are already known by experimental means. This is a relatively small database, and the lack of availability of suitable template structures is the main limitation to the use of comparative modeling at the present time. This situation may change in the coming years. Large-scale projects aimed to create high throughput structure determination, under the name of *structural genomics*, are now underway. It is likely that suitable template structures will become available for most protein sequences within the next 10-15 years.

When template structures have been obtained it is necessary to align their sequences with the target sequence using a multiple alignment tool. From a user's point of view, the alignment process is the most crucial step in comparative modeling. If the alignment is good, then a good model will result; if it is poor, then the model will also be poor. When the target and template sequences are closely related, with high percentage identities (for example 70% and above) then automated alignment methods usually produce good alignments. Nevertheless, even when percentage identity is very high, the alignment should be inspected for any obvious alignment errors. For example, it is advantageous to check that conserved key structural and functional residues are aligned. If there are key structural cysteines, glycines or prolines, then these should be aligned with residues of corresponding types in all the sequences.

When target and template sequences are more weakly related, with lower percentage identities, then alignment errors are more common and more manual checking is necessary. It may be very difficult to identify the correct alignment, and this can result in poor predictions. There are now many software package that do completely automated comparative modeling, with no manual user input. These function well when target and templates are closely related, but should be treated with caution when relationships are more distant. Always check that the alignment used was acceptable from a structural and functional perspective.

8.8.3 The Process

The various software packages carry out the comparative modeling process in slightly different ways, and what we describe here are generic features of the process that are common to all methods. The process begins by analysis of the template structures. If there is more than one such structure, an average structure is often calculated by superposing the structures in three dimensions and calculating average atomic positions. The contributions of the various templates in this process are often weighted according to their degree of similarity to the target sequence, with more similar sequences more strongly weighted. In this process a framework of template atomic positions is calculated.

Next, using the alignment of the target sequence to the templates, the structure is divided into two distinct types of region, the structural core and the nonconserved loops. The core can be considered to comprise mainly the secondary structure elements (helices and strands) of the templates, within which each residue is aligned with a residue from the target sequence. The nonconserved loops are mainly surface loops for which, according to the alignment, target and template sequences have different numbers of residues (i.e. there are gaps in the alignment). Structure prediction for the core is easy, because the backbone atomic positions from the averaged template structure can be used directly as backbone atomic positions for the predicted (model) structure of the target sequence.

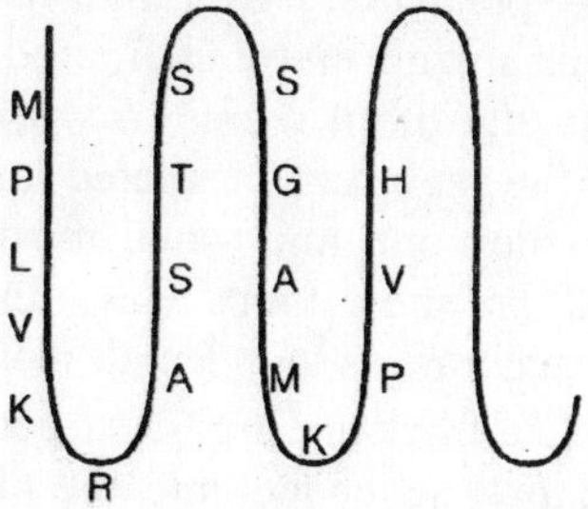

Sequence of known structure:

MPLVKRASTSSGAMKPVH...

Schematic of known 3D structure (template)

Sequence of unknown structure (target):

MPILKRGTSTSYGAMRPIY...

Aligned with sequence of known structure:

```
MPILKRGTSTSYGAMRPIY
MPLVKRASTSS_GAMKPVH
```

Predict the 3D structure of the target sequence by replacing the old residues on the known structure with those from the target sequence made equivalent by the alignment.

Need to position new side chain atoms.

Small structural changes where gaps are found in the alignment (loop modelling, dotted circle).

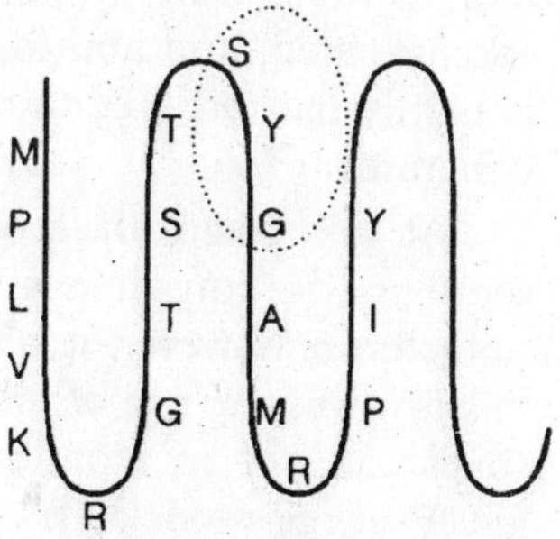

Predicted 3D structure

Figure 8.13. Comparative modeling.

Within nonconserved loops, structure prediction is more difficult, and more sophisticated structure prediction methods have to be used. The simplest method of loop structure prediction is to use the *spare parts* algorithm. This method makes use of a database of known loop structures from other proteins that are not necessarily similar in sequence to the target. For each loop needed, a spare part loop is found in this database that fits into the gap in the modeled structure, and this is taken as the predicted structure for the loop region.

There are more sophisticated loop prediction algorithms, involving more complex calculation and minimization of estimated loop energies, but the simple spare parts algorithm seems to work well in practice. Prediction accuracy in the region of predicted loops is generally much lower than it is in the core, and while most loop prediction algorithms perform well on short loops (less than six residues) they become increasingly inaccurate as loop length increases.

When the steps above are complete there are predicted positions for all the backbone atoms of the target sequence, and the process turns to the question of *side-chain positions*. Some side-chain identities are conserved between target and template structures, and the structural information from the templates can be used to place these side-chains in the predicted structure. However, not all side-chains are conserved and the positions of the atoms in these need to be predicted. This is usually done by choosing from a library of allowed side-chain structures (sometimes called a *side-chain rotamer library*), so that the atoms of the side-chains fill the available space in the protein interior without being too close (clashing) with any other protein atoms. The algorithms used to do this are often quite sophisticated, and aim to produce a protein hydrophobic core that is tightly packed, like those observed in experimentally determined structures.

At this stage, the preliminary structural model for the target sequence is complete and many predictions stop at this point. Sometimes, however, it is useful to refine the model further. Typically this involves the use of energy minimization software that can make small changes to atomic positions in order to produce a slightly lower energy model. This can be advantageous, for instance to move any atoms that were left too close in proximity after the modeling process above, but the advantage in terms of the accuracy of the model is debatable.

8.8.4 Accuracy

We have already commented that the accuracy of comparative models is determined almost entirely by the accuracy of the alignment. When alignments are good, which they typically are when the sequences are very closely related, then very accurate models are possible. Accuracy is typically measured in terms of root mean square deviation (RMSD) between the a carbon positions in the predicted structure and the actual structure of the target sequence RMSDs of less than 1.0 Å represent very good predictions; this degree of difference between the two structures is similar to the degrees of difference between two separate experimental determinations of the same protein structure.

When the percentage sequence identity between template structures and target sequence exceeds 70% it is reasonable to expect that the model should be accurate to an RMSD of less than 2-3 Å, even using completely automated methods. When the percentage identity drops below 40% then getting a good alignment, often with manual intervention, becomes crucial and automated methods can fail very badly.

It is of course not possible to know how accurate a model is without an independent experimental determination of the actual structure, but the above guidelines should .be borne in mind. It is also possible to check some aspects of the predicted structure; for instance, software for checking of stereochemical parameters of the structure will identify any unusual bond lengths or angles that might indicate potential problems in the prediction.

Predicted structures should be viewed in general as quite low-resolution models. For applications where a very accurate structure is required, for example detailed structure-based drug design, they are probably too low in accuracy, particularly in respect of nonconserved loops and side-chain positions. Nevertheless, if it is possible to construct a structure prediction by comparative modeling then this prediction is likely to be more accurate than a prediction by any other method.

8.8.5 Availability

Comparative modeling is available on the World Wide Web (WWW) at the SWISS-MODEL site, and using the SWISS-PDBVIEWER software (http://www.expasy.ch/swissmod/SWISS-MODEL.html). It is also available in several commercial software packages.

8.9 SECONDARY STRUCTURE PREDICTION

8.9.1 What is Secondary Structure Prediction?

The previous topic introduced comparative modeling as the method of choice for protein structure prediction. When suitably related template structures do not exist for a particular target sequence, secondary structure prediction is a viable alternative. Unlike comparative modeling, it does not produce a full atom model of the tertiary structure, but rather just provides a prediction of the secondary structure state of each residue, either *helical*, *strand* or *extended*, or *coil*. The predictions are sometimes known as *three-state predictions*.

Most methods of secondary structure prediction have been trained on the database of known structures. Two early methods were those of Chou and Fasman, and GOR (Gamier-Osguthorpe-Robson, after its inventors). Although these methods work in different ways, the former being rule based and the latter using information-theoretical ideas, they both make use of the idea of *secondary structure propensity*. The amino acids seem to have preferences for certain secondary structure states. For instance, glutamic acid has a strong preference for the helical secondary structure, and valine a strong preference to be in strands. Glycine and proline have lower than average propensity for both types of regular secondary structure, reflecting a tendency to be found in loops, and the some hydrophobic amino acids (e.g. phenylalanine) have strong preferences for both types of secondary structures reflecting their tendency to make up the structural core. However, none of the preferences are particularly strong, and all amino acids are found in each type of secondary structure quite often. This means that predictions of secondary structure cannot be based on single residues.

In the Chou-Fasman method, a helix is predicted if, in a run of six residues, four are helix favouring, and the average value of the helix propensity is greater than 1.0 and greater than the average strand propensity. Such a helix is extended along the sequence until a proline is encountered (these are known to break helices) or a run of four residues with helical propensity less than 1.0 is found. A strand is predicted if, in a run of five residues, three are strand favouring, and the average value of the strand propensity is greater than 1.04 and greater than the average helix propensity. Such a strand is extended along the sequence until a run of four residues with strand propensity less than 1.0 is found. This is a simple rule-

Table 8.2. Helical and strand propensities of the amino acids.

Amino acid	*Helical (α) propensity*	*Strand (β) propensity*
GLU	1.59	0.52
ALA	1.41	0.72
LEU	1.34	1.22
MET	1.30	1.14
GLN	1.27	0.98
LYS	1.23	0.69
ARG	1.21	0.84
HIS	1.05	0.80
VAL	0.90	1.87
ILE	1.09	1.67
TYR	0.74	1.45
CYS	0.66	1.40
TRP	1.02	1.35
PHE	1.16	1.33
THR	0.76	1.17
GLY	0.43	0.58
ASN	0.76	0.48
PRO	0.34	0.31
SER	0.57	0.96
ASP	0.99	0.39

based method dependent on finding runs of residues with preference for one type of secondary structure. The GOR method is different, considering the information carried by a residue about its own secondary structure, in combination with the information carried by other residues in a local window of eight residues on either side in the sequence of the residue concerned.

The accuracy of these early methods based on the local amino acid composition of single sequences was fairly low, with often less than 60% of residues being predicted in the correct secondary structure state. This should be viewed in light of the fact that for proteins containing roughly equal proportions d helix, strand and coil then the accuracy of random predictions would be 33% (i.e. the correct state would be predicted for one in every three residues).

8.9.2 Multiple Sequence Information

In the late 1980s, it was realized that the evolutionary information contained in multiple sequence alignments could significantly improve secondary structure predictions. Several factors contribute to this increase in accuracy, but they all relate to the patterns of conservation revealed in the multiple alignment. First, a residue with a high propensity for a particular secondary structure in one sequence may have occurred by chance, but if it is part of a conserved column in which all the residues have high propensity for that type of secondary structure then this provides much more predictive evidence. Second, multiple alignments can reveal more subtle patterns of conservation. For instance, a large proportion of a-helices in globular proteins are amphipathic.

These helices have hydrophobic and hydrophilic faces and are associated with periodic patterns of hydrophobic and hydrophilic residues in the sequence. The appearance of conserved pattern of this type is therefore highly predictive of α helical structure. Finally, as already discussed, insertions and deletions are much more likely in surface loops (coil), so regions of the alignment where these are common are more likely to be in the coil state. The use of evolutionary information from multiple alignments improved the accuracy of secondary structure prediction methods significantly, resulting in methods capable of an accuracy of around 66% at the time.

8.9.3 Accuracy of State-of-the-art Methods

Since the late 1980s, we have seen an explosion in the amount of data in the sequence and structure databases that has increased the amount of evolutionary and structural information available to secondary structure prediction algorithms. This, coupled with improvements to the algorithms themselves, has led to much increased accuracy. Examples of state-of-the-art prediction methods are PHD (Rost and Sander), DSC (King and Sternberg), PREDATOR (Frishman and Argos), NNSSP (Salamov and Solevyev) and PSI-PRED (Jones). These methods use a variety of techniques to make their predictions, many of which are based in the fields of artificial intelligence or machine learning. PHD and PSI-PRED make use of artificial neural networks (simple computational models of the neural networks in the brain), and DSC is based on a machine learning method known as linear discriminant analysis. NNSSP (Nearest Neighbour Secondary Structure Prediction) makes predictions based on locating the most similar sequence segments in the database of known structures. The accuracy of all these methods is above 70%, and, depending on the test set of predictioru chosen, accuracies of above 75% have been achieved. This prediction was carried out using a protein of very recently determined structure, and the actual secondary structure from that experiment is shown for comparison with the predicted structure. The overall accuracy in this case is higher than average (83%). It should be noted that many of the prediction errors-are residues on the ends of helices or strands and that the secondary structure of these residues may not be well defined in the protein: it can be very difficult to decide whether a residue at the end of a helix is part of the helix or not.

One caveat that must always be born in mind is that all secondary structure prediction methods have been trained using structures from the database of known protein structures. This set of proteins is biased for various reasons; some very common proteins (for instance integral membrane proteins) are under-represented. The prediction methods can be expected to work better for proteins that are in some sense similar to the proteins in this database, and less well for others. It makes no sense therefore to apply standard secondary structure prediction methods to integral membrane proteins.

8.9.4 Prediction of Transmembrane Segments

As already stated that the structure of integral membrane proteins as consisting of segments (usually helical) spanning the lipid membrane, connected by loops lying outside the membrane. The membrane-spanning segments tend to contain a high proportion of hydrophobic residues and are often more than 20 residues in length, corresponding to six to seven helical turns for trans-membrane helices. Such relatively long runs of predominantly hydrophobic residues are seldom found in water-soluble globular proteins. This means that it is often possible to predict, based on the runs of hydrophobic residues, whether or not a protein is an integral membrane protein, and if so, where the membrane-spanning segments are in the sequence.

Tools have been developed to predict transmembrane segments, including TMPred, TMHMM and TopPred. Many of these tools also predict the membrane-spanning topology. This is a prediction of the orientation of the helices with respect to the membrane. For instance and i→o (in to out) helim have their N-termini inside whatever cell or organelle the membrane bounds, and their C-termini outside. Clearly, a consistent prediction of topology musl have i→o helices followed by o→i helices and *vice versa.* Predictions of the membrane-spanning topology are generally less reliable than predictions membrane-spanning segments irrespective of orientation.

Table 8.3. Prediction of transmembrane helical segments for human rhodopsin (a G protein-coupled receptor).

Helix	*Predicted position*	*Actual position*	*Predicted orientation*	*Actual orientation*
1	37-63	37-61	o→i	o→i
2	74-99	74-98	i→o	i→o
3	115-140	114-133	o→i	o→i
4	153-175	153-176	i→o	i→o
5	203-221	203-230	o→i	o→i
6	253-276	253-276	i→o	i→o
7	286-309	285-309	o→i	o→i

An example of a prediction of membrane-spanning segments is shown in *Table* 8.3, where predictions have been made for the much-studied G protein-coupled receptor rhodopsin. In this case the predictions are quite accurate, with all seven transmembrane segments predicted in the correct sequence. This is of course not always the case, and it is quite common for some real membrane spanning segments not to be predicted (false negatives), or for some non membrane-spanning segments to be predicted as membrane spanning (false positives).

False negatives often occur in ion channel proteins, where some of the membrane-spanning helices have hydrophilic sides, which cluster together to make a hydrophilic path through the membrane along which charged ions can pass. The extra hydrophilic residues in these helices often mean that the) evade detection by methods searching for hydrophobic residues. False positive are often found in secreted proteins that contain an N-terminal signal peptide. Such signal peptides tend to be hydrophobic, and this sometimes leads to their confusion with membrane-spanning segments.

8.9.5 Availability of Tools

Most of the secondary structure and transmembrane segment prediction tools mentioned in this topic are available from the ExPASy WWW site (http://www.expasy.ch), under proteomics tools.

8.10 ADVANCE PROTEIN STRUCTURE PREDICTION AND PREDICTION STRATEGIES

8.10.1 Fold Recognition

As already discussed the evolution of protein structures and established that similarity in sequence is sufficient to guarantee similarity in structure. The method of comparative modeling is based on this fact. However, we also established that during evolution, structure is much more strongly conserved than sequence. There are many cases of distantly related proteins with the same structure but whose level of sequence similarity is well below the 25% needed to guarantee structural similarity and the possibility of comparative modeling. Fold-recognition methods are about detecting these distant relationships, and separating them from chance sequence similarities not associated with a shared fold. They operate by searching through a library of known protein structures (called a *fold library*) and finding the one most compatible with query sequence whose structure is to be predicted. An alternative name for fold recognition is *threading*.

The product of fold recognition is usually an alignment between the query sequence and one or more distantly related sequences of known structure. Fold recognition can therefore be viewed as an extension of the comparative modeling method to very distant relationships. Once the alignment between the sequence and the distantly related known structures has been obtained, a full three-dimensional structure of the protein to be predicted can be obtained using the usual methods of comparative modeling. Because relatively few protein structures are known, most new sequences cannot have their structures predicted by ordinary comparative modeling. Fold recognition is valuable because it has the potential to extend significantly the number of protein sequences whose structures can be predicted.

As already stated that distinction between homologous proteins which share the same fold because they have evolved by divergent evolution from a common ancestor, and analogous proteins which share the same fold perhaps for physicochemical reasons with little evidence for common ancestry. Fold-recognition methods were originally intended to detect both analogs and homologs of the query sequence, but current evidence suggests that they can detect distant homology but not analogy.

Fold-recognition methods employ a variety of approaches, most often mixing the similarity in sequence detected by amino acid substitution matrices with further structural information. For instance the 3D-PSSM (*three-dimensional Position Specific Scoring Matrix*) uses fold library structures described in terms of ordinary one-dimensional sequence profiles generated by PSI-BLAST, and also 3D profiles. The information within the 3D profile includes secondary structure and solvation potential. The solvation potential takes account of the tendency of hydrophobic amino acids to occupy structural positions that are not accessible to solvent (buried in the hydrophobic core). The secondary structure component measures the degree of similarity between the predicted secondary structure of the query sequence, using one of the methods described already, and the secondary structure of the fold library member. There is very good evidence that the inclusion of structural information like secondary structure and solvation results in methods that are better able to detect distant homology than the most sensitive sequence only methods (for instance PSI-BLAST), and therefore that these are useful in structure prediction. Several fold-recognition methods operating in related ways to 3D-PSSM are linked from the text WWW site. The URL for 3D-PSSM is http://www.sbg.bio.ic.ac.uk/~3dpssm/.

One problem with fold-recognition methods is that while they are often able to recognize a distant homology to a known fold, they sometimes do not provide particularly accurate alignments to that fold. This situation is improving as the methods develop, but it must always be borne in mind that alignments may not be very accurate. Inaccurate alignments lead to poor structural predictions.

8.10.2 *Ab initio* Prediction

Ab initio methods were already discussed as methods that attempt to predict protein structures from first principles using theories from the physical sciences like statistical thermodynamics and quantum mechanics. We will say little more about them here, except to set them in context of the methods we have already described. There is no doubt that first principles predictions are very attractive from an intellectual viewpoint. However, these methods are still not sufficiently well developed to be of real use in practical protein structure prediction. Proteins in their natural solvation environments are very large systems, and are currently beyond the scope of accurate calculation with accepted theories. This is not to say that research in *ab* initio methods is uninteresting, indeed to understand the process of protein folding in this way would be a huge scientific achievement; but, from the point of view of the practicing molecular biologist or biochemist, the methods are currently of little use by comparison with the methods of comparative modeling, secondary structure prediction and fold recognition described previously.

8.10.3 Strategies

We close this section by outlining a strategy for protein structure prediction for a new query sequence of unknown structure. The first step should be perhaps to identify any features of the sequence that might affect the strategy to be adopted. The sequence should be examined for potential membrane spanning segments that might indicate the presence of an integral membrane domain. There are now some integral membrane proteins of known structure, but not many, and structure prediction for such domains remains difficult. Other sequence features that should be identified could be regions of low compositional complexity, since few of these appear in the database of known structures and therefore structure prediction might be inaccurate in such regions. The presence of coiled coils could be tested. Finally an analysis of the sequence by a tool like Interpro might reveal the presence of known domains and perhaps the overall domain structure of the sequence. This could be supplemented by a PSI-BLAST search, which might reveal other related sequences and the sub-sequences (domains) where they are related. If the protein is multi-domain, it would make sense to predict each domain separately, if the positions of domains in the sequence could be defined.

The most accurate and comprehensive structure prediction method is comparative modeling, so the first prediction step of the strategy is to see if this is possible. This requires the identification of sequences of known structure with sufficient similarity to the query sequence, and can be carried out at the SWISSMODEL WWW site. If this search is successful, then a comparative model can be built and there is no need for any further action. All other structure prediction methods are less accurate and less useful than comparative modeling. Even to make a secondary structure prediction is not useful, because the secondary structure of the comparative model itself is likely to be more accurate than any other secondary structure prediction method.

Comparative modeling is only possible for a minority of new protein sequences. When it is not possible, the next logical step is secondary structure prediction. This can be applied to any sequence, but of course is more accurate for globular protein domains and likely to be less accurate for integral membrane domains and low complexity regions. This will give a prediction of helix, strand or coil for each residue, which might be of use, for example in mutagenesis studies. Following secondary structure prediction, fold-recognition methods might give an idea of how the secondary structures pack into the tertiary fold, but these methods should be applied with some caution. It must be remembered that fold recognition detects very distant relationships where even structural divergence can be significant, and that 3D structure predictions from this method are generally much less accurate than those from standard comparative modeling.

9

Industrial Application

9.1 BIOINFORMATICS AND DRUG DISCOVERY

9.1.1 Drugs

A *drug* is a molecule that interacts with a *target* in the body (a biological molecule, usually a protein) thus triggering a physiological effect. Drugs can be beneficial or harmful depending on that effect. Drugs for the treatment of diseases interact with targets that contribute to the disease and produce positive effects. The disease target may be endogenous (a protein synthesized by the individual to whom the drug is administered) or, in the case of infectious diseases, may be produced by a pathogenic organism. Drugs act by either stimulating or blocking the activity of the target protein.

9.1.2 Principles of Drug Development

The development of a new drug is a complex, lengthy and expensive process. Current estimates place the total development costs of a new drug (including the writing off of false starts, clinical trials and tests required by regulatory authorities) somewhere in the region of 800 million dollars. The viability of the exercise requires a company to be able to launch one or two new drugs per year whereas the actual figures are typically 0.4 -0.8 new drugs per year per company. This is one reason for the consolidation (through mergers and takeovers) of the large pharmaceutical companies.

Drug development begins with the identification of a potentially suitable disease target, a process called *target identification*. In the past, target identification was based largely on medical need. Nowadays, with many useful therapies already on the market or in development, target identification depends not only on medical need but also on factors such as the success of existing therapies, the activity of competing drug companies and commercial opportunities.

The next stage of the process is *target validation*, which involves extensive testing of the target molecule's therapeutic potential. Validation may include the creation of animal disease models and the analysis of gene and protein expression data under various conditions. A valid target must contribute to a human disease and have a high *therapeutic index*. This means that, should an appropriate drug be found, a significant therapeutic gain must be predicted through the use of such as drug. Once a target has been validated, the search begins for drugs that interact with that target. This process is called *lead discovery*, and involves the search for *lead compounds*, that is, substances with some but not all of the desired, biological activity of the ideal drug.

Lead optimization involves the modification of lead compounds to produce derivatives, *candidate drugs* (*CDs*), with better therapeutic profiles. For example, increased specificity for the target would be desirable because this would result in reduced side-effects. The CDs are then assessed for quality, taking into account factors such as ease of synthesis and formulation. Development candidates passing this test may then be registered as an *investigational new drug* and submitted for *clinical trials*. This is the lengthiest and most expensive part of the drug development process, which is why most, projects are abandoned before this stage. Clinical trials are designed to determine safety and tolerance levels in humans, and to discover how the drug is metabolized. Trials are divided into several stages:

1. Preclinical phase: animal studies.
2. Phase I: normal (healthy) human volunteers.
3. Phase II: evaluation of safety and efficacy in patients, and selection of dose regimen.
4. Phase III: large patient number study with placebo or comparator; at this stage regulatory approval is sought and a commercial launch decision taken.
5. Phase IV: long-term monitoring for adverse reactions reported by pharmacists and doctors.

9.1.3 Bioinformatics in Drug Development

Over the last 10 years, drug development has benefited greatly from advances in biology and chemistry, predominantly in three areas: genomics/proteomics, combinatorial' chemistry and high-throughout screening. Due to these advances, the scale and complexity of target identification/validation and lead discovery / optimization experiments has increased dramatically, resulting in huge amounts of data that need to be stored and analyzed. The term *pharmainformatics* is often used to describe the mix of biology, chemistry, mathematics and information technology required for data handling in the pharmaceutical industry.

Genomics and proteomics have revolutionized the way target molecules are identified and validated. Traditionally, drug targets have been characterized on an individual basis and lead compounds sought with specific clinical effects. With the advent of genomics, particularly the availability of the entire human genome sequence and its annotations, thousands of potential new targets can now be identified by sequence, structure and function. Bioinformatics is important not only because of its role in the analysis of sequences and structures, but also in the development of algorithms for the modeling of target protein interactions with drug molecules. This allows *rational drug design*, in which protein structural data is used to predict the type of ligands that will interact with a given target, and therefore form the basis of lead discovery.

Another new development based on genomics is the emerging scientific field of *pharmacogenomics*. This is the study of the association between genomic data and drug response patterns. Some individuals with a disease may benefit from a drug while others receive little or no benefit or may experience drug-related side-effects. Such individual variations may be due to underlying variation in the structure of the target molecule, or differences in drug metabolism, both of which depend on genotype. Variations in drug response patterns are responsible, in part, for the high failure rates of new drug candidates at the clinical trials stage. *Genomic medicine* is the concept of treating patients with drugs tailored to their genotype, which can be rapidly screened by searching for individual changes in the nucleotide sequence, known as *single nucleotide polymorphisms* (*SNPs*, pronounced 'snips'). The way in which lead compounds are identified has also changed significantly over the last 10 years. Traditionally, there are four sources of lead compounds:

1. Lead compounds may be discovered through *chance observations* (the most famous example of this was the discovery of penicillin, which has given rise to many of today's antibiotics).

2. They may also be identified on the basis of *side-effects* caused by drugs developed for alternative treatments (e.g. the tranquilizer chlorpromazine' was discovered following its experimental use as an antihistamine).
3. Lead compounds may be derived from *traditional medicines*, for example quinine from cinchona bark.
4. Lead compounds may be *natural substrates* or *ligands* (e.g. epinephrine (adrenaline) which has given rise to analogs such as salbutamol, which is used to treat asthma).

More recently, these traditional approaches have been joined by more systematic methods for lead identification. These methods are based on *high-throughput screening* (*HTS*) in which lead discovery is accelerated through the use of highly parallel assay formats, such as 96-well plates. In turn, this requires the assembly of large chemical libraries for testing. This has been made possible by *combinatorial chemistry* approaches, in which large numbers of different compounds can be made by pooling and dividing materials between reaction steps. Bioinformatics is important in all these developments: for the storage, tracking and analysis of HTS data, and for the design and assessment of chemical libraries.

The impact of genomics, proteomics, combinatorial chemistry and HTS on drug development is difficult to judge, simply because the time from target identification to the marketing of a successful new drug is so long. However, these advances increase the number of compounds that can be screened by several orders of magnitude and should also make it easier to choose between a potentially successful drug and a likely nonstarter. More than 99% of all drug development projects are abandoned before the new drug is marketed and only a third of marketed drugs actually recuperate the costs of research and development. Although these high-throughout methods will increase the research and development budgets of the pharmaceutical companies significantly, there should also be a good return in the form of marketable novel drugs.

9.2 PHARMAINFORMATICS RESOURCES

9.2.1 Pharmainformatics

The combination of biology, chemistry, mathematics and information technology required for data processing and analysis in the pharmaceutical industry is sometimes termed *pharmainformatics*. The need for pharmainformatics in drug development reflecting the impact of genomics, proteornics, combinatorial chemistry and high-throughput screening (HTS) on drug discovery. In this topic, we consider specific applications of pharmainformatics in more detail and discuss available database and software resources.

9.2.2 Chemical Diversity

HTS in drug discovery depends on the availability of diverse chemical libraries, such as those generated by combinatorial chemistry, since these maximize the chances of finding molecules that interact with a particular target protein. It is difficult to quantify *chemical diversity* but attempts have been made based on the concept of 'chemical space'. Essentially, *chemical space* encompasses molecules with all possible chemical properties in all possible molecular positions. A diverse library would have broad coverage of chemical space, leaving no gaps and having no clusters of similar molecules. In practice, library diversity is quantified using measures that compare the properties of different molecules based on *descriptors* such as atomic position, charge and potential to form different types of chemical bond. Two molecules can be compared using the *Tanimoto coefficient* (T_C), which evaluates the similarity of fragments of each molecule. The coefficient is calculated as shown-below where *a* is the number of fragment-based descriptors in compound A, *b* is the number of fragment-based descriptors in compound B and c is the number of shared fragment-based descriptors:

$$T_C = c/(a + b - c)$$

Thus, for identical molecules, $T_C = 1$, while for molecules with no descriptors in common, $T_C = 0$. In a chernical library of ideal diversity, most pair-wise comparisons would generate a Tanimoto coefficient near to zero.

Diverse libraries are very useful for lead discovery when little is known about the binding specificity of the target protein. However, it is more often the case that some form of sequence or structural information is available for the target, and this can be used in the design of *focused libraries* that concentrate on one region of chemical space. For example, if the sequence of a particular target protein is known, then database homology searching will often find a related protein whose structure has been solved and whose interactions with smal! molecules have been characterized. In such cases, it is possible to design a chemical library based on a particular *molecular scaffold*, which preserves a framework of sites present in a known ligand, but which can be *decorated* with diverse functional groups. Some of these groups may have previously been shown to be important for drug binding (such sites are known as *pharmacophores*).

Tools and resources are available for the design of combinatorial libraries and the assessment of chemical diversity. For example, a program called *Selector*, available from Tripos, allows the user to design very diverse libraries or libraries focused on a particular molecular skeleton. *Chern-X*, developed by the Oxford Molecular Group, allows the chemical diversity in a collection of compounds to be measured and identifies all the pharmacophores. Another Tripos program, *CombiLibMaker*, allows a virtual combinatorial library to be generated and used to screen a virtual target. Similar resources are available from MDL Information Systems.

9.2.3 Computational Screening

Laboratory-based screening experiments are very expensive, so it is beneficial to generate as much information about potential drug/target interactions as possible before such an experiment is undertaken. The *computational screening* of chemical databases, using a target molecule of known structure, is one way in which such information can be obtained. The structure of the target molecule may be known from X-ray crystallography or nuclear magnetic resonance spectroscopy data. Alternatively, the solved structure of a close homolog may be used, or the structure may be predicted using a threading algorithm. If the structure of a target protein is known, algorithms can be used to identify potential interacting ligands based on goodness of fit, therefore allowing *rational drug design*. Several well-known drugs have been developed in this manner, including Relenza and Captopril.

Many *docking algorithms* have been developed which attempt to fit small molecules into binding sites using information on steric constraints and bond energies. One of the most established docking algorithms is *Autodock*. Initially the software comprised three modules, Autotors, Autogrid and Autodoc. Autotors dealt with ligand co-ordinates and the definition of bonds that were axes of free rotation, Autogrid constructed a three-dimensional grid of interaction energies, and Autodock performed the docking simulation itself. Another widely used program is *DOCK*, in which the arrangement of atoms at the binding site is converted into a set of spheres called *site points*. The distances between the spheres are used to calculate the exact dimensions of the binding site, and this is compared to a database of chemical compounds. Matches between the binding site and a potential ligand are given a confidence score, and ligands are then ranked according to their total scores. This modeling approach has the disadvantage that the binding site and every potential ligand are considered to be stiff and inflexible. Other algorithms can incorporate flexibility into the structures. A more recent development called *CombiDOCK* considers each potential ligand as a scaffold decorated with functional groups. Only spheres on the scaffold are initially used in the docking prediction and then individual functional groups are tested using a variety of bond torsions. Finally, the structure is *bumped* (checked to make sure none of the positions predicted for individual functional groups overlap) before a final score is presented.

Chemical databases can be screened not only with a binding site (searching for complementary molecular interactions) but also with another ligand (searching for identical molecular interactions). Several available algorithms can compare two-dimensional or three-dimensional structures and build a profile of similar molecules. This approach is important, for example, if a ligand has been shown to interact with a protein but has negative side-effects, or if a structurally distinct ligand is required in a drug development project to avoid intellectual property issues. In each case, a molecule of similar shape with similar chemical properties is required, but with a different structure.

9.2.4 Functional Genomics

Functional genomics incorporates any branch of biology dealing with the high throughput functional annotation of genes, that is, the determination of gene functions on a genomic scale. Functional genomics is set to have a massive impact on the pharmaceutical industry because it provides a fast track to target identification and validation. Thus far, there are no examples of drugs that have been developed solely on the basis of functional genomics because the field is new, but the impact has already been felt in the early stages of drug development and it is likely that the output of the industry, in terms of newly registered drugs, will increase dramatically in the next 5-10 years.

Bioinformatics and functional genomics go hand in hand because the very nature of functional genoinics means that large amounts of data are generated. Some aspects of functional genomics are covered elsewhere in this book, specifically the prediction of protein function by sequence comparison and structural analysis, the analysis of gene and protein expression data and the analysis of protein interaction data. All of these play a major role in the pharmaceutical industry. One aspect of functional genomics that is of particular relevance to drug development, and that has not been discussed elsewhere, is *genome-wide mutagenesis*, the deliberate creation of mutants for each and every gene in the genome.

In the pharmaceutical industry, the mouse is the most important animal because its physiology and biochemistry are so similar to those of humans. Drugs can be tested on healthy mice and on those in which the disease in question has been artificially induced (such an animal is known as a *disease model*). In many cases, disease models are produced by genetic manipulation, generating either a *transgenic mouse* (a mouse with additional genetic material) or a *knockout mouse* (a mouse in which a particular gene has been mutated, disrupted or removed by homologous recombination). In the past, disease models have been created on an individual basis by targeting particular genes. Over 15000 different transgenic and knockout mouse strains have been generated in individual projects, providing the material for many potential disease models. These strains are listed in a number of dedicated databases, the most comprehensive of which are *T-BASE* and the *BioMedNet knockout mouse database*. Disease models can also be generated by *random mutagenesis* rather than the creation of specific strains. Large-scale mutagenesis of the mouse has been carried out using a chemical mutagen called ENU (*ethylnitrosourea*). Over 40000 mutagenized mouse strains have been generated and scored for phenotypes relevant to drug development, including allergy, immunology, clinical chemistry, nociception (response to pain), dysmorphology (abnormal development) and behaviour. Another method for the generation of random mutants is *insertional mutagenesis*, where a DNA construct is integrated randomly into the genome and may disrupt a gene. Currently, two organizations, the *German Gene Trap Consortium (GGTC)* and the US biotechnology company *Lexicon Genetics* are carrying out projects to generate genome wide mutant libraries in which every possible mouse gene is interrupted. Mutagenesis is carried out in *embryonic stem cells*, which are derived from the early mouse embryo and have the potential to generate all cell types. Banks of mutant ES cells can be stored indefinitely at low temperatures. At the appropriate time, the cells can be thawed and injected into normal mouse embryos where they colonize and contribute to the germ line, allowing the production of transgenic mice in the subsequent generation. So far more than 20000 mutant cell lines have been produced in the two projects.

Bioinformatics has many roles to play in the management of genome-wide mutagenesis projects. The cataloging of mutant strains and the tracking of their behaviour under many experimental conditions is one example. The creation of databases allowing mouse strains to be searched by gene, phenotype and clinical relevance is another. Random mutagenesis projects involve a further issue, which is the identification of the interrupted genes. The GGTC and Lexicon Genetics maintain databases of *flanking sequence tags*, short DNA sequences I derived from the flanks of their integrated DNA constructs, which can be searched against GenBank to help identify the gene that is interrupted.

It should be noted that, although the mouse is the most relevant model organism to humans, genes and proteins do tend to be highly conserved across the animal kingdom. The functional similarity between gene products in humans and other animals is so conserved that some biotechnology and pharmaceutical companies are using a factory-style screening approach in which thousands of mutant strains of a simple organism, for example the nematode worm *Caenorhabditis elegans,* are tested with a panel of potential drugs to identify interactions that rescue or otherwise influence the mutant phenotype. The advantage of this approach is the highly parallel assay format and the fact that these simple organisms have complete genome sequences with highly informative annotations. The mutant gene in C. *elegans* can thus be identified rapidly and used to identify homologous genes in the human genome that may act as novel drug targets.

9.2.5 Pharmacogenomics

Pharmacogenomics is the study of association between genome data and drug response patterns. Pharmacogenomics is an important new development because it offers the promise of *personalized medicine*, that is, drugs targeted to an individual's genome. This should help to avoid the clinical problem of *adverse drug reaction*, a severe negative reaction to drug administration that is thought to kill as many as 100 000 people a year. There is currently no easy method for predicting an individual's response to drug treatment, so pharmaceutical companies are restricted to the development of drugs that are suitable for all patients, which severely limits their scope. In the future, pharmacogenornics may allow the development of more specialized and targeted drugs.

Bioinformatics plays a vital role in pharmacogenomics because it allows data on genetic variation to be compared meaningfully with data on drug response patterns. Correlations, generated by clustering similar data, can divide patients into discrete *response groups*. For example, there are four recognized response groups to the drug nortriptyline: poor metabolizers, intermediate metabolizers, extensive metabolizers and ultra-rapid metabolizers. These phenotypic groups reflect allelic variants of the *CYP2D6* locus, which encodes an enzyme of the cytochrome P450 oxidase family. Patients in each group are given different doses of nortriptyline commensurate with their metabolic characteristics.

The analysis of variation among the human population is facilitated by databases that catalog mutations and expression patterns. Expression databases are include the National Center for Biotechnology Information (NCBI) Gene Expression Omnibus (GEO). The NCBI holds a second database, *dbSNP*, which lists *human single nucleotide polymorphisms* (*SNPs*). These are individual nucleotide changes which, when they occur in proteincoding genes, may be associated with many different phenotypes including variations in the absorbance or clearance of drugs. The rationale behind the collection of SNPs is the creation of a large data set that will allow profiles of SNPs to be built up and matched against particular drug response patterns. In the future, this will allow individual patients to be 'typed' for their likely response to drug administration, and SNP data is likely to be incorporated into the early stages of drug development (target validation) as well as forming an important part of the clinical trials stage.

10

Genetic Algorithm

10.1 COMPOSITION ALIGNMENT

Most algorithms which characterize DNA functional sites concentrate on identifying *position-specific patterns,* such as consensus patterns or weight-matrix-based profiles. Each is a string $P = p_1 p_2 .. p_k$ in which p_i is specified as either a single character or a choice of characters. The data for a pattern are typically collected from multiple known occurrences of the functional site. Well known examples include the TATA box, a frequent component of the eukaryotic gene promoter, where transcription from DNA to RNA starts of the eukaryotic gene promoter, where transcription from DNA to RNA starts and the weight matrices used to describe transcription binding sits in the TRANSFAC database. Often, position-specific pattern shave low selectivity. That is, when used to search for unknown occurrences, they yield many non-functional sites. This, in part, is due to the nature of the patterns which are frequently short and degenerate (more than on letter choice at some positions) making random matching a relatively common event.

A obvious reason is that functional sites contain multiple features working synergistically and accurate recognition requires identifying several of those features properties which contribute to function *are no reducible to position-specific patterns* and thus are not easily incorporated into current search techniques. An example is double helical strand separation, a necessary antecedent for several important DNA functions. Rather than exhibiting position specific properties, the propensity for strand separation appears to be spread, in a complex way, over thousands of nucleotides. Programs have been developed to calculate regions of conjectured low energy requirement for strand separation and these programs have had some success in identifying sites which are functionally dependent on this process. In this paper, we develop a new approach for DNA sequence analysis which employs the mechanisms of string matching and string comparison yet avoids the overdependence of these methods on position-by-position matching. We initiate the study of what we call the *composition property*, which manifests as similarity in *composition*, rather than in position-specific patterns.

Definition: *Composition* is a vector quantity describing the frequency of occurrence of each alphabet letter in a particular string. Let S be a string over Σ. Then $C(S) = (f_{\sigma 1}, f_{\sigma 2}, ..., f_{\sigma|\Sigma|})$ is the composition of S, where for $\sigma_i \in \Sigma$, $f_{\sigma i}$ is the fraction of the characters in S that are σ_i. Note that the *order* of letters in S is irrelevant as it has no effect on the composition of S. Two strings S and T have a *composition match* if their lengths are equal and $C(S) = C(T)$. There is an accumulating body of evidence that variation in composition along the DNA strands contributes to function. There are a number of important DNA features, at both large and small scales, whose unifying characteristic is composition has including.

10.1.1 Protein Binding Regions

Within these domains, tens of nucleotides long, dinucleotide, or base-step composition, can contribute to DNA flexibility, allowing the helix to change physical conformation, a common property of protein-DNA interactions.

10.1.1.1 Isochores

These multi-megabase regions of genomic sequence are specifically GC-rich or GC-poor. GC-rich isochores exhibit greater gene density. Human ALU and L1 retrotransposons appear preferentially in isochores with composition that approaches their own.

10.1.1.2 CpG islands

These regions of several hundred nucleotides are rich in the dinucleotide CpG which is generally underrepresented (relative to overall GC content) in eukaryotic genomes. The level of methylation of the cystine (*C*) in these dinucleotide clusters has been associated with gene expression in nearby genes.

The spring board for our study of the composition property is a new alignment algorithm which we call *composition alignment*, or more whimsically, *scrambled alignment*. Standard sequence alignment is based on single character matching. In composition alignment, we extend the matching concept to substrings which have the same composition i.e., composition matching. This allows us to identify subsequences that share regions of similar composition. More specifically, composition alignment is a pairing of substring of *exactly matching* composition separated by insertions, deletions or mismatches.

As an example let

$X = AACGTCTTTGAGGTC$

$Y = AGCCTGACTGCCTA$

One possible composition alignment for X and Y is

```
AACGTCTTTGAGCTC
| | <->   |<--->
AGCCTGACT-GCCTA
```

where symbols between the letters are used to indicate single character matches (|) and substring composition matches (<--->). Note that composition matches (Either single or multicharacter) can occur consecutively in an alignment.

An idea related to composition alignment is the "Swap" operation in string comparison, first discussed by Lowrance and Wagner. A swap or transposition is the exchange of two letters that are side-by-side. Matching two letters and their transposition is equivalent to a composition match for substrings of length two. Lowrance and Wagner gave a $O(nm)$ time algorithm for alignment including the swap operation. More recently, Amir et al. gave an algorithm for finding all swapped matching of a pattern of length m in a text of length n with time complexity $O(nm^{1/3} \log m \log \sigma)$. This was later improved to $O(n \log m \log \sigma)$ and followed by an algorithm to count the number of swaps in each swapped match within the same time complexity.

10.1.2 Problem Description

The problem we address in the following:

10.1.2.1 Composition alignment

Given. Two sequences, S of length m, and T of length n, over an alphabet Σ, and a scoring function $cm(s, t)$ for the score of a composition match between substrings s and t.

Find. The best scoring alignment (global or local) of S with T such that the allowed scoring options include composition match between substrings of S and T as well as the standard options of single character match, single character mismatch, insertion and deletion.

For the remainder of this paper, we will assume (1) that the alphabet Σ is fixed and (2) that the similarity scoring is used. The latter means that matches are given positive weight; mismatches, insertion and deletions are given negative weight; and the best alignment has the highest score.

10.1.3 Composition Alignment Algorithm

Our composition alignment algorithm is similar to standard alignment algorithms and is computed using dynamic programming. A single step in filling the dynamic programming array, W, can be analyzed in the following way. Given two sequences $X = x_1 \ldots\ldots x_n$ and $Y = y_1 \ldots y_m$, the best composition alignment of the two prefix strings $X[1, i] = x_1 \ldots x_i$ and $Y[1, j] = y_1 \ldots y_j$ ends in one of the following four ways.

New:

1. A *composition match* between suffix of length l, $1 \leq l \leq \min(i, j, limit)$, i.e. $x_{i-l+1} \ldots x_i$ and $y_{j-l+1} \ldots y_j$, where *limit* is an upper bound on the length of a substring that can participate in a composition match.

Standard:

1. A *mismatch* between x_i and y_j.
2. A *deletion* of x_i.
3. A *deletion* of y_j.

The score in cell (i, j) is the maximum obtained by these four possible (global alignment) or the maximum among these four and a score of zero (local alignment). Note that the score of alternative 1 is

$$W[i - l, j - l] + cm\ (x_{i-l+1} \ldots x_i, y_{j-l+1} \ldots y_j),$$

where W is the alignment score matrix and cm() is the score function for composition matches. The time complexity of the algorithms is $O(nmZ)$ where Z is the time required, per (i, j) pair, to find the best length l in alternative 1 and compute its score. In the next section, we show how to precompute the *length* of the shortest composition match for every (i, j) pair in constant time per pair.

10.1.3.1 Finding composition matches

Our goal here is to find the length *l*, of the shortest suffixes of the strings X[1, *i*] and Y[1, *j*] which have a composition match, and to do this for every (*i*, *j*) pair, in constant time per pair. For example, if *X* = *AACGTCTTTGAGCTC* and *Y* = *AGCCTGACT GCCTA*, then for the pair (4, 8) the shortest suffices that have a composition match are X[2,4] = *ACG* and Y[6, 8] = *GAC* each with a length of 3. To find these matches, we use composition difference.

Definition. Composition difference is a vector quantity for two strings. Let *x* and *y* be strings over an alphabet Σ. Then *CD* (*x*, *y*) = $(c_{\sigma 1}, ..., c_{\sigma|\Sigma|})$ is the composition difference of *x* and *y*, where $c_{\sigma r} = c^{y}_{\sigma r} - c^{y}_{\sigma r}$ is the difference between the number of times σ_r occurs in *x* ($c^{x}_{\sigma r}$) and the number of times it occurs in *y* ($c^{y}_{\sigma r}$). For example, let Σ = {0, 1}, *x* = 010111010001000 and *y* = 010001101110111. Then *CD* (*x*, *y*) = (3, –3) because *x* has three more zeros and three fewer ones than *y*. We compute composition match lengths for one diagonal of the alignment matrix at a time.

Each diagonal is defined by a value *d*, $-n \le d \le m$ and contains the index pairs (*i*, *j*) such that *j* – *i* = *d*. The substrings processed in a diagonal for *d* ≥ 0 are *X* [1, *k*] and *Y*[1 + *d*, *k* + *d*] and for *d* ≤ 0 are *X*[1 – *d*, *k* – *d*] and *Y*[*i*, *k*] for all values of *k* = 1, ... *min*(*m*, *n*) such that both strings are non-null. To illustrate let *S* = 010111010001000 and *T* = 010001101110111. We show how to process diagonal zero. Figure 10.1A is a plot of the composition differences for successive prefixes of *S* and *T*. Since the components of any composition difference vector must sum to zero, we plot only the excess of ones in *S* relative to *T* (the second number in the composition difference vector). The *key observation* is that two identical composition differences at prefix lengths *g* and *h* with *g* < *h*, indicate a composition match between the sub-string of *S* and *T* from position *g* + 1 to position *h*, i.e., of length *h* – *g*.

As shown in the plot of our example, prefix lengths 3 and 10 both have composition difference (0, 0) so there is a composition match between the substrings *S* and *T*. Similarly prefix lengths 4, 7 and 9 each have composition difference (–1, 2), so there is a composition match for the substring pairs (*S*, *T*, *S*, *T*), and (*S*, *T*). Note that prefix length 11 has composition difference (1, –1) which is not shared by any prefix length shorter than 11. Therefore, three is *no composition* match for the prefix ending at position 11 in these two strings. To identify composition matches, we compute the ordered pairs (*prefix length, composition difference*) and then sort these using composition difference as the key. This is done with *radix sort* across the σ values and *counting sort* within each σ value.

Notice that both sorts are stable, *i.e.* they do not rearrange elements with the same key. Shortest composition matches are determined by scanning the sorted list to find *adjacent* elements with the same key (composition difference). The difference between prefix lengths is then stored in an array $ML(i, j)$. In our example, since prefix lengths 10 and 3 are adjacent with the same composition difference, we store 7 (= 10 – 3) as the shortest composition match for prefixes of S and T of length 10.

Time complexity

The sorting is linear in the number of elements when the alphabet is fixed. For a single diagonal of the alignment matrix, the number of elements is the diagonal length. Over all diagonals, the number of elements is $(n + 1)(m + 1)$. Therefore the time complexity to preprocess the strings to find all shortest composition match lengths is $O(nm)$.

10.1.3.2 Scoring function

The overall complexity of our composition alignment algorithm depends on the complexity of computing the best length l for a composition match. This in turn depends on the scoring function, $cm()$, for composition matches. Below, we discuss several scoring functions we have tested.

Function based on match length

In this group, the score of a composition match depends on the length, k of the match we have tested

- Function 1: $cm(k) = ck$
- Function 2: $cm(k) = ck$
- Function 3: $cm(k) = c\log(k + 1)$

where c is a constant. These function are additive or subadditive, meaning that $cm(i + j) \leq cm(i) + cm(j)$. Function 1 treats matches of different lengths equally Function 2 and 3 (subadditive) give less weight, per character, to long composition matches, than to short composition matches.

A convenient property of additive or subadditive functions is that, when computing alternative 1 of the alignment score, for any

(i, j) pair, it is sufficient to find the length of the *shortest* suffices of $X(1, i)$ and $Y[1, j]$ which have a composition match.

Lemma. For an index pair (i, j) let $l = l_1 < l_2 <...< l_k, 1 \leq 1\ min(i, j)$, be the lengths for which there is a composition match between the suffixes $X[i - l + 1, i]$ and $Y[j - l + 1, j]$. Then, the score for the best alignment which ends in a composition match between suffixes of $X[1, i]$ and $Y[1, j]$ is equal to the score when the suffixes have length l_1. That is $\forall h, 2 \leq h \leq k, W(i - l_1, j - l_1) + cm(l_1) \geq W(i - l_h, j - l_h) + cm(l_h)$.

Proof. Assume by way of contradiction that there is an $l_h > l_l$ such that

$$W(i - l_h, j - l_h) + cm(l_h) > W(i - l_l, j - l_l) + cm(l_1).$$

Let $l_h = l' + l_1$. Then

$$W(i - l_h, j - l_h) + cm(l' + l_1) > W(i - l_1, j - l_1) + cm(l_1)$$

$$W(i - l_h, j - l_h) + cm(l' + l_1) - cm(l_1) > W(i - l_1, j - l_1)$$

but by additivity or subadditivity,

$$cm(l') \geq cm(l' + l_1) - cm(l_1)$$

so

$$W(i - l_h, j - l_h) + cm(l') > W(i - l_1, j - l_1)$$

which is a contradiction because $W(i - l_h, j - l_h)$ is assumed to be optimal including the possibility that the alignment which yields this score ends with the composition match of length l'.

This means that breaking up a long composition match into shorter matches (if possible) will leave the score the same or increases the score. The alignment shown in the introduction contains a 4 character composition match which is broken into a single character match and a 3 character match.

Theorem. Composition alignment with an additive or sub-additive composition match scoring functions has time complexity $O(nm)$.

Functions based on substring composition

Here, the score of a composition match depends not just on length, but on the composition of the matching substrings. We have tested:

- Function 4: $cm(x, y, k) = ck \cdot H(C, B)$

where x and y are substring with common composition C, k is their length, and c is a constant. $H(C, B)$ is the relative entropy of composition C given a background composition B. *Relative entropy* is defined as

$$H(C,B) = -\sum_{\sigma \in \Sigma} f_\sigma \log(f_\sigma / b_\sigma)$$

where f_σ is a frequency in the composition vector C and b_σ is the corresponding frequency in the background composition B. This function can only be used if for every non-zero f_σ in C there is a corresponding non-zero b_σ in B, else there will be a divide-by-zero problem. In our studies this has not arisen because we use the overall-frequency of letters in the sequences to be aligned as the background. If divide-by-zero is possible, H can be replaced with the unweighted Jensen Shannon divergence. We have not yet tested this function extensively. Function 4 favors composition matches where the substring differ significantly from the background. This could, for example, be a long repetition of a single letter, assuming the background is relatively balanced. This function is not additive or subadditive, so finding the shortest composition match for any (i, j) pair does not always yield the optimal match length.

Since a longer match may yield a higher alignment score, we must test all substring match lengths per (i, j) pair. This can easily be done in time linear in the number of match length by stepping through the *ML* array which stores the shortest match lengths, but requires at most $\min(m, n)$ tests per (i, j) pair. In practice though, the *limit* parameter explained below restricts the number of tests to at most *limit* per (i, j) pair.

Theorem. Composition alignment using function 4 for composition match scoring and the limit parameter has time complexity $O(nm \cdot limit)$.

Retrieving the alignment

After the alignment score array W has been computed, the optimal alignment is retrieved by tracing back as in standard alignment. When a score $W[i, j]$ is derived from a composition match, we need to reference the length of the matching substring. For Functions 1, 2 and 3, this is done by querying the ML$[i, j]$ value. For Function 4, we must store the optimal match lengths separately as the score are being computed and then refer to these values when tracing back. In either case, retrieving the alignment requires $O(n + m)$ time.

10.1.4 Alphabet Size and the Limit Parameter

The study of alignment using similarity scoring has shown that for ungapped local alignments of randomly generated sequences, the parameter space for match and mismatch weights is divided into logarithmic and linear regions. In the logarithmic region, the parameters produce alignment scores proportional to the logarithm of sequence lengths whereas in the linear region, the scores are directly proportional to the sequence lengths. It is generally accepted that weight combinations which fall within the logarithmic region are useful for detecting *biologically related* sequences, whereas those in the linear region do not distinguish between related and unrelated sequences.

The same general features have been observed in gapped local alignments where gap weight is an additional parameter. The rubric for determining if parameters fall within the logarithmic or linear regions is to look at the expected score per aligned letter pair (ungapped alignments) or the expected global alignment score (gapped alignments. In either case, if the expected score is negative, and assuming that positive scores are possible, then the parameters fall within the logarithmic region. Here we are primarily interested in how the *limit* parameter, the length of the longest allowed composition match, fits into this framework.

When *limit* = 1, composition alignment is equivalent to standard alignment. When *limit* = 2, any pair of adjacent letters in one sequence is allowed to match its transposition in the other sequence. This corresponds to the swap operation mentioned earlier. For *limit* = 3, both scrambled triplets and transposed doublets are allowed the amount of matching. If too much matching occurs, then the average score will be positive and the alignments will not be meaningful.

10.1.4.1 Expected fraction of matching characters in alignments

We have examined both ungapped and gapped composition alignments to determine the expected fraction of aligned letter pairs that are counted as matches.

10.1.4.1.1 Ungapped alignments

Suppose we have a binary alphabet and we examine ungapped aligned strings of length 2 where the characters are generated iid with probability 0.5. Under single character matching, the expected fraction of characters counted as matching is 0.5, i.e. on average half the characters will be counted as matches. When we allow composition matches with *limit* = 2, the expected number of characters counted as matches increased to 0.625. For aligned strings

of length 2 or 3 can match as long as they have the same composition), the expected fraction of matches is 0.6875.

As the sequence length grows calculating the fraction of matches becomes complicated, so we turn to simulation. When the string length reaches 10, and we allow composition matching with *limit* = 10, the fraction of matches is above 0.82. For the four letter DNA alphabet, when the letters are generated iid with probability 0.25, the fraction of characters matching grows similarly but more slowly. For the sixteen letter dinucleotide alphabet, the probability grows until reaching an apparent asymtotic upper bound around 0.075. For dinucleotides, we first generate an iid DNA sequence and then convert it to a dinucleotide sequence.

Table 10.1. Fraction of characters counted as matching in randomly generated ungapped alignments where *limit* equals alignment length, for three alphabets.

Sequence length	1	2	3	4	5	6	7	8	9	10
Binary (%)	50.0	62.5	68.75	72.7	75.6	77.3	78.9	80.3	81.3	82.4
DNA (%)	25.0	30.0	32.3	35.3	37.5	39.7	40.7	42.4	43.3	44.2
Dinucleotide (%)	6.2	6.5	6.7	6.9	7.1	7.3	7.3	7.3	7.4	7.5

Notice that consecutive letters in the dinucleotide sequence are not independent (i.e. if the first dinucleotide is AC, then the next must start with a C). To investigate the fraction of matches in longer sequences where the *limit* is smaller than the sequence length, we use global composition alignment to count the matches. Here, insertions and deletions are not allowed, all matches are weighted 1 and all mismatches weighted zero.

The fraction matching in DNA sequence is nearly 45% with limit = 5 and reaches 50% limit = 9. For dinucleotides, the fraction levels off at 7.78% for limit $\geq$ 20. For *local, ungapped* composition alignments with DNA sequences using alignment parameters (composition match constant, single character match, mismatch) = (1, 1, – 1), alignment scores grow in proportion to the log of the sequence length for limit $\leq$ 5. For *limit* between 6 and 10, growth is proportional to the square root of the sequence length. Note that with these alignment parameters, the expected score of an aligned pair is negative until *limit* = 9. Thus negative expected score per aligned letter pair is an inaccurate predictor of logarithmic score growth for ungapped composition alignments.

10.1.4.1.2 Gapped alignments

For gapped alignments, we use simulations with actual parameter values and composition match scoring functions because the interaction of these determines the number of gapped positions. The results are useful for several purposes:

1. To define values for the limit parameter that fall within the logarithmic region.
2. To test for concurrence between (1) the change from negative to positive average global alignment score and (2) the change from logarithmic to linear behaviour in the local alignment score. This is not obvious because the assumptions that underly the theory for alignment scores do not include scrambled alignment.
3. To estimate local alignment score distributions so that the composition alignment algorithm can be used to search for statistically significant alignments in real biological sequences.

We used DNA sequences generated iid with probability 0.25 for each letter and two sets of alignment parameters (composition match constant, single character match, mismatch, indel), (2, 2, –3, –5) and (2, 2, –7, –7). Here we summarize some of the more important results.

Fucntion 1. Local alignment scores grow in proportion to the logarithm of sequence length for $limit \leq 3$. At $limit = 4$, scores are proportional to the square root of sequence length. Note though, that the average global alignment score does not become positive until $limit = 5$. These results indicate that function 1 should be used with *limit* set to 3 and not higher. Also, positive or negative expected global score is inaccurate as a predictor of the parameter values that yield logarithmic growth in local alignment scores.

Functions 2 and 3. Local alignment scores for function 2 are logarithmically related to sequence length below $limit = 10$. Average global scores do not become positive with *limit* as high as 50. Function 3 behaves similarly. Again, average global score is an inaccurate predictor of local alignment score behaviour.

Function 4. Local alignment scores are proportional to the log of the sequence lengths upto $limit = 50$.

10.1.5 Examples

We tested our composition alignment algorithm on a set of 1796 human promoter sequences from the Eukaryotic Promoter Database (EDP) maintained by the Bioinformatics Group of the Swiss

Institute for Experimental Cancer Research. The database contains a collection of roughly 3000 annotated non-redundant eukaryotic RNA polymerase II promoters for which the transcription start site has been experimentally determined. Each sequence is 600 bases long and consists of 500 bases upstream and 100 base downstream of the transcription initiation point.

The sequences are non-redundant, as selected from the database, which means that no two share greater than 50% sequence identity. The sequences were aligned pairwise using composition match scoring function 1 with alignment parameters (composition match constant, single character match, mismatch, indel) of (2, 2, –7, –7). This produced a score *W*. Each pair was also aligned with a standard alignment algorithm using the same parameter values, producing a score S. Those pairs for which (1) *W* was above the statistical significance cutoff score for composition alignment for a set of sequences this large (as determined by simulation) and (2) $W \geq 3 \cdot S$ were retained.

The second criterion was used to exclude composition alignments that scored highly because they were redetecting good standard alignments. Two high scoring alignments are shown here. The first example was obtained with the promoter pair EP27006 and EP73975. The standard local alignment which is not statistically significant is shown for comparison. The composition alignment is characterized by high GC content (89%), an enrichment over the background frequency of these sequences (78%). The number of CpG dinucleotides found in the aligned regions is more than expected given either the background composition of these subsequences or the entire sequences.

This suggests that the aligned regions are part of CpG islands which are defined as being 200 bp subsequences with a C + G content exceeding 50% and a ratio of observed CpG to expected CpG in excess of 0.6 CpG islands are known to occur in the 5' region of many genes. This alignment is typical of many obtained with the promoter set. A second example involves promoters EP73298 and EP11149. Again the standard local alignment score for this pair is not statistically significant.

An interesting feature of the composition alignment is the change in composition of the subsequences from left to right. The left two thirds is GC rich with C dominant, and a single A (0.01, 0.61, 0.30, 0.08). The situation changes at the right with is G dominant with the fraction of As, and Cs equivalent (0.19, 0.6, 0.56, 0.09). The fraction of T is roughly the same throughout. Notice that the background composition for these sequences is typical, GC rich with the complementary nucleotides balanced: (0.15, 0.36, 0.34, 0.15).

10.1.6 Conclusion

We define a new type of alignment problem, *composition alignment* which extends the matching concept to substrings of equal length and the same nucleotide composition. We give an algorithm for composition alignment which has time complexity $O(nm)$ for a fixed alphabet when the composition match scoring function is additive or subadditive.

The time complexity is $O(nm \cdot limit)$ for a relative entropy scoring function where *limit* is an upper bound on the length of the substring that can match. We explore how *limit* fits into the framework of the logarithmic and linear regions of alignment parameter space. When computing gapped alignments, using our function 1, *limit* should be set to 3. For functions 2 and 3 *limit* should be under 10 and for function 4, *limit* can be as high as 50. We give two examples of composition for human RNA polymerase II promoters where the composition alignment scores are statistically significant even though there is no detectable similarity with standard alignment.

10.2 OPTIMAL MULTIPLE PARSIMONY ALIGNMENT WITH AFFINE GAP COST USING A PHYLOGENETIC TREE

An algorithms for pairwise parsimony alignment of biological sequences was devised by Needleman and Wunsch in 1970. The minimum cost of an alignment is found by a recursive algorithm in the prefixes of the two sequences. Once the matrix of optimal prefix alignment costs has been found, the optimal alignment can be found by going back through the matrix (backtracking). The cost of gaps in this algorithm is linear function of the gap length.

The original pairwise alignment algorithm had a time and memory complexity of $O(L^2)$ (for sequence of length L). In 1975, however, Hirschberg described an algorithm for pairwise alignment that reduced the memory complexity to $O(L)$, while only increasing time consumption by a factor of two. In 1982, Gotoh published an algorithm for pairwise sequence alignment that can use a gap cost which is an affine function of the gap length, rather than linear. This was done in a way resembling the Needleman-Wunsch algorithm while keeping track of three different types of residue configurations of the last alignment step: (1) residues in both sequences, (2) a residue in sequence one aligned to a gap in other, and (3) a residue in sequence two aligned to a gap in sequence one. The use of an affine gap cost improves the alignments of biological sequences significantly. At close to the same time, Sankoff and Cedergren described an algorithm for multiple parsimony alignment over a tree with linear gap cost. Other methods have used the sum of pairs scores for multiple alignments, rather than scores based on tree. This scoring scheme does not have an evolutionary explanation, which gives it some drawbacks.

There has also been various approximation methods proposed for multiple alignment. This work gives an algorithm for optimal multiple parsimony alignment with affine gap cost for any number of sequences related by a tree. It combines ideas from pairwise alignment with affine gap cost, and ideas from multiple alignment with linear gap cost. The memory reducing technique of Hirschberg is also applied. The resulting algorithm has a memory complexity of $O\ (7.442^N L^{N-1})$ and a time complexity of $O(16.81^N L^N)$ (For N sequences). These complexities should be compared to $O(\mathrm{L}^{\mathrm{N}-1})$ and $O(2^N L^N)$, respectively, for the linear gap cost method. Some examples are given to show the utility of the algorithm. Due to the large time and memory complexities, it is not practical to use for more than three sequences or may be four short sequences. The method could, however, prove useful in a number of other situations, such as alignment quality evaluation and pair-HMM based probabilistic alignment, particularly when using sampling based approaches.

10.2.1 Problem

10.2.1.1 Pairwise parsimony alignment

Gaps in pairwise alignments are the result of insertions and deletions (indels) of sequences pieces. The cost of indels in the affine gap cost setting is based on a gap introduction cost ($g_i > 0$) and a gap extension cost ($g_e < g_i$), giving a total cost of $g_i + (l - 1)g_e$ for an indel of length l. For a sequence alphabet Σ of size n, a symmetric $n \times n$ matrix, $\{m_{ij} \geq 0\}$, gives the cost of matches for all possible pairs of residues. The cost of matching two identical residues is zero. Here is an example:

```
G C G G G T -        Cost m_GC + m_CC + g_i + g_e + m_GG + m_TT + g_i
C C - - G T A                 = m_GC + 2g_i + g_e
```

$$\text{Cost } m_{GC} + m_{CC} + g_i + g_e + m_{GG} + m_{TT} + g_i = m_{GC} + 2g_i + g_e$$

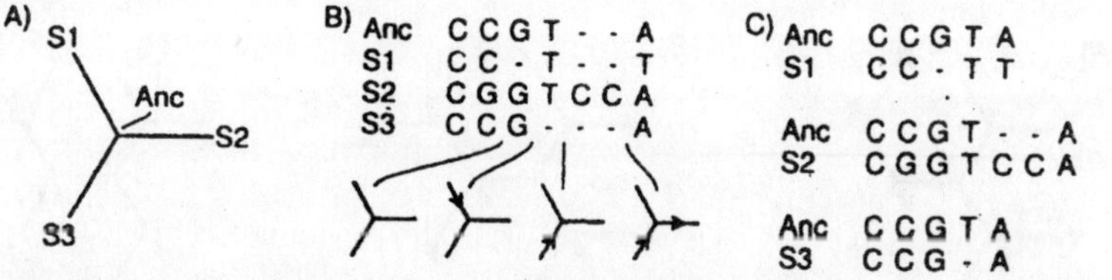

Figure 10.1. A–The phylogenetic tree relating the three sequences S1, S2 and S3, and their common ancestor. B–The multiple alignment of the four sequences and indel configurations at various points. C–The implied pairwise alignments.

Notice that the total indel cost is independent of the nature of the residues, given the alignment. The optimal alignment of two sequences is the one that minimizes the total cost.

10.2.1.2 Multiple parsimony alignment

Assume that *N* sequences are given, as well as an unrooted binary tree relating them. The *N* sequences are located at the leaves of the tree, while the ancestral sequences are at the internal vertices. Assume for a moment that the ancestral sequences are known, so an alignment of all these sequences implies a pairwise alignment of the two sequences at the ends of each edge in the tree. For these pairwise alignments, columns with gaps in both sequences are removed. Now, the price of the whole alignment is the sum of the costs of all the implied pairwise alignments of neighbouring sequences. If the ancestral sequences are not given, the cost of the multiple alignment of the *N* sequences is the minimum cost over all possible ancestral sequences and all possible alignments of the sequences. An optimal multiple alignment of the *N* sequences is one with minimal cost.

10.2.1.3 Objectives

Given *N* sequences, a phylogenetic tree relating them, a matrix of match costs, a gap introduction cost, and a gap extension cost, we seek the optimal multiple parsimony alignment of the sequences.

10.2.2 Algorithm

10.2.2.1 Residue configurations

For a given tree, label each vertex with either a residue (≠) or a gap (–), and denote that a residue configuration for the tree. Call a residue configuration *acceptable* if (1) the vertices labeled by residues form a connected graph and (2) at least one leaf has a residue. A residue configuration will be used to define what occurs in a single column of a multiple alignment.

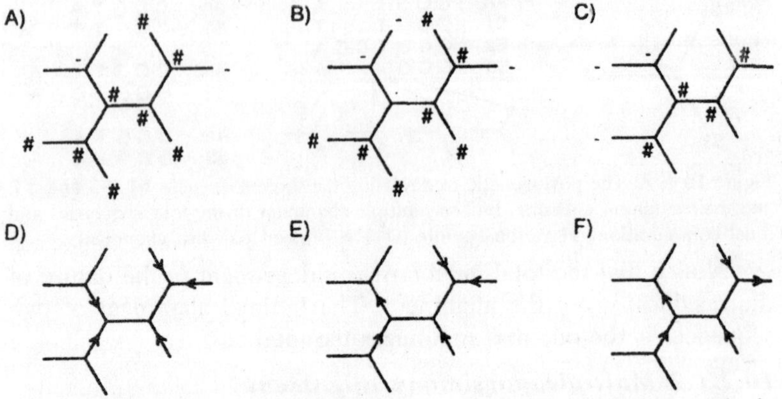

Figure 10.2. The three top trees show various residue configurations, while the bottom three shows indel configuration.

When aligning two residues, the biological interpretation is that they are related to each other, which is the reasoning behind condition one of an acceptable residue configuration. A column with no residues in any of the sequences being aligned can not be part of an optimal alignment (since $g_i > 0$), leading to condition two for an acceptable configuration.

10.2.2.2 Indel configurations

Given a tree, we can choose to label each edge in one of three ways: (1) an arrow pointing from one end to other, (2) an arrow pointing the opposite direction, and (3) just by a line. We will call such a labeling an indel configuration for the tree. For an indel configuration on a tree, a vertex, *a*, is isolated from another, *b* if the path from *a* to *b* has an arrow pointing opposite to the direction of the path.

An indel configuration is *acceptable* if it satisfies the following constraints: (1) no vertex is isolated from all the leave and (2) this still holds if the direction of any single arrow is reversed. The indel configurations are used for keeping track of ongoing indels in a multiple alignment. There is no immediate biological reasoning for the unacceptable indel configurations, but it turns out that they can never be part of an optimal alignment.

10.2.2.3 Recursion

Define a prefix alignment of the *N* sequences as the optimal alignment of certain prefixes of the sequences. An *N* dimensional integer vector *ν* defines the number of residues from each sequence that is included in the prefix alignment. The algorithm starts with the empty prefix alignment (no residues from any sequences) and builds it up column by column, while keeping track of the minimal cost. Using this approach, the optimal cost of the entire alignment can be found, and by backtracking through the costs and optimal alignment may also be found. Notice that there could be more than optimal alignment. For a prefix alignment, the indel cost of an additional column will depend on which pairwise indels have been initiated, since the gap cost is affine.

The indel configurations are used to keep track of ongoing indels between two sequences that are neighbours on tree. An arrow indicates that an indel was present between the sequences at the ends of the corresponding edge in the last column involving any of these two sequences. The arrow points in the direction of the sequence that had a residue and away from the one with a gap. A line with no arrow indicates a match in the last column involving the two sequences. Because of the dependence on past indels, we will keep track of each possible ending indel configuration for the prefix alignments. Let $E_v^{(d')}$ denote the optimal cost of the prefix alignment of the sequences corresponding to the sequence vector v and indel configuration d.

The addition of a column in the alignment may change the indel configuration depending on which matches are made and which gaps are introduced and extended. For a given indel configuration, let us say that we apply a residue configuration when adding a column with the corresponding residues to give a new indel configuration. The change in indel configuration is given by one of the following three possibilities. (1). An edge with residues at both ends changes the indel configuration to a line at that edge (At no cost). (2) An edge with a residue at one end, and a gap at the other, changes the indel configuration to an arrow pointing toward

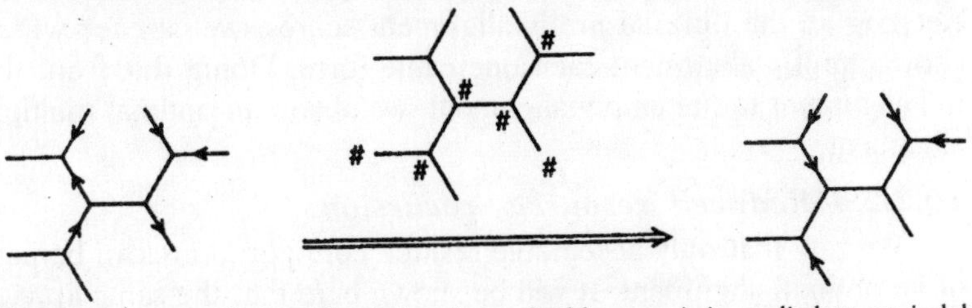

Figure 10.3. A residue configuration (over double arrow) is applied to an indel configuration (lef) giving a new indel configuration (right).

the residue (at a cost of g_e or g_i if the same arrow was there already or not, respectively) (3) An edge with gaps at both ends does not change the indel configuration (and has no cost). Thus the gap cost of applying a residue configuration to an indel configuration is the gap cost of the new alignment column, given the existing prefix alignment. For a vector, *e*, with coordinate values of one or zero for each sequence (including ancestors), the corresponding residue configuration is the one with residues for ones and gaps for zeros.

For an indel configuration *d'*, applying the residues configuration corresponding to the vector yields another indel configuration, *d*. We write this as $d' \xrightarrow{e} d$. The gap cost of applying the residue configuration is written $G_e^{(d')}$. When going from prefix alignment $v - \hat{e}$ to v (with $\hat{e}$ denoting the vector with only the *N* entries of *e* corresponding to the leaves), the residues of the sequences are known, and the optimal ancestral residues are found by a post order tree traversal giving the minimal residue match cost. The match cost is written as $M_{(v-\hat{e})\rightarrow v}$, which is not dependent on which ancestral sequences has residues, since matching two identical residues is free ($m_{ii} = 0$).

For the alphabet, Σ, the match costs can be found for all possible residue configurations before starting the recursion, which will have no effect on the time or memory complexity for large *L*. Each acceptable residue configuration defines a possible column in the alignment, and thus a step from one prefix alignment to the next. We can write the algorithm for finding the optimal cost as this recursion:

$$E_v^{(d)} = \min_{e,d' \xrightarrow{e} d} \{E_{v-\hat{e}}^{(d')} + G_e^{(d')} + M_{v-\hat{e}-v}\} \quad \text{...(1)}$$

with $E_0^{(d')}=0$ for the indel configuration, *d* where all edges are labeled with a line, and $E_0^{(d')} = \infty$ for all other indel configurations, *d'*. Keeping all the optimal prefix alignment scores, we can see which shorter prefix alignment each one came form. Doing this from the full alignment to the empty alignment, we obtain an optimal multiple alignment.

10.2.2.4 Reduced resources recursion

We saw that only acceptable residue configurations can be part of an optimal alignment. It will be shown here that the same applies to indel configurations. Assume that we have an indel configuration where some vertices are isolated from all the leave. Let a be the last of these vertices to have a residue (or one of the last if there is more than one), and consider the column in the prefix alignment where that occurred. Since only acceptable residue configurations are used, this column must have had a residue at a leaf and there would be a connected set of vertices with residues between a and the leaf. Since this is the case, a can not be isolated from that leaf immediately after this column.

The only way to isolated a from the leaf is by having a later column with a residue at α and none at the leaf, which is a contradiction. Now, assume that we have an ideal configuration where no vertices are isolated from all leaves, but the reversal of an arrow at the edge α isolates some vertices from all leaves. The last column involving residues at either end of α must have had a residue at one end and a gap at the other end.

The vertex, a, that had the gap must become isolated upon the reversal of the arrow at α. Thus the last column, C_i, with a residue at a could not have and a residue at any of the leaves on the same side of the tree relative to α. Let Cj be the last column before C_i, where a leaf on this side of the tree had a residue. The prefix alignment of all sequences (including ancestral), leading to this indel configuration, could be changed to have a gap at a, and all other vertices on that side of the tree, in the columns C_{j+1} to C_i. This would not change the alignment of the leaf sequences. Further more, the cost of this prefix alignment would be lower, since the match cost would be the same and it has less gap introductions or less gap extensions, or both. This means that the assumed indel configuration could never be part of an optimal alignment if $g_e >$ 0. If $g_e = 0$, both configurations could be optimal, so the assumed indel configuration is again not necessary to ensure finding an optimal alignment. With this, it has been shown that only acceptable indel configuration need to be taken into account in the recursion. Furthermore, for certain indel configurations only a subset of the acceptable residue configurations (the ones leading to acceptable indel configurations) needs to the applied.

10.2.2.5 Hirschberg's memory reduction

In (1), each prefix alignment cost $E_v^{(d)}$, only depends on the costs for v', where any coordinate is the same or one lower than v. Denote a row of costs as a set of costs where the coordinate of v corresponding to sequence one is the same. To find the cost of the optimal alignment, we only need to keep the present and the previous row of costs in memory. This reduces the memory complexity by a factor of L. With this we can not, however, find the optimal alignment since we lost the information we need to backtrack. Using a backwards recursion for finding the prices from the full alignment to shorter prefixes, an optimal prefix in the middle of the alignment (relative to the first sequence) can be chosen and the alignment problem can be split into two problems of smaller size. The two problems together will take at most half as much time as the original problem, so the total time will be no more than a factor $1 + {}^1/_2 + {}^1/_4 + ... = 2$ longer. With this the memory complexity has been reduced by a factor of L, while the time complexity has only increased by a factor of two.

10.2.3 Algorithm Analysis

Here, it is assumed that the tree is binary. For trees with higher vertex orders, the time and memory limits will still hold.

10.2.3.1 Memory complexity

The memory complexity of the algorithm is in order of L^{N-1} times the number of acceptable indel configurations for N sequences. The indel configurations for a rooted tree can be represented by a three dimensional vector, f. The first coordinate is the number of indel configurations where the root is not isolated from all leaves and reversing a single arrow does not change that. The second coordinate is the number of indel configurations where the root is not isolated from all leaves, but reversing a single arrow changes that. The third coordinate is the number of indel configurations where the root is isolated from all leaves.

The configurations for a rooted tree can be calculated from the two rooted subtrees (a and b) connected to the root as:

$$f_{ab} = [f_a A f_b^T \; f_a B f_b^T \; f_a C f_b^T],$$

where:

$$A = \begin{bmatrix} 6 & 3 & 1 \\ 3 & 1 & 0 \\ 1 & 0 & 0 \end{bmatrix} B = \begin{bmatrix} 2 & 2 & 1 \\ 2 & 2 & 1 \\ 1 & 1 & 0 \end{bmatrix} C = \begin{bmatrix} 1 & 1 & 1 \\ 1 & 1 & 1 \\ 1 & 1 & 1 \end{bmatrix}$$

These relationships can be found by connecting of the two subtrees to a vertex (The new root) with al nine combinations of labels for the two new edges.

Two rooted trees (a and b) can be connected by a single edge to give an unrooted tree. There are three configurations for the connecting edge, but only some of these lead to acceptable indel configurations for the whole unrooted tree, depending on the configurations of a and b. By going through these possibilities the total number of acceptable indel configurations for the unrooted tree can be determined as:

$$f_a \begin{bmatrix} 3 & 2 & 1 \\ 2 & 1 & 0 \\ 1 & 0 & 0 \end{bmatrix} f_b^T, \qquad \text{...(2)}$$

where f_a and f_b are the configurations for the two subtrees.

Define a linear tree as one where any vertex is at most one edge away from a leaf. Here a proof will be sketched that for a given number of leaves, the linear tree has the highest number of acceptable indel configurations. Consider the following rearrangement of a tree, where T_1 and T_2 represent subtrees with at least two leaves and dashed lines represent a linear connecting tree:

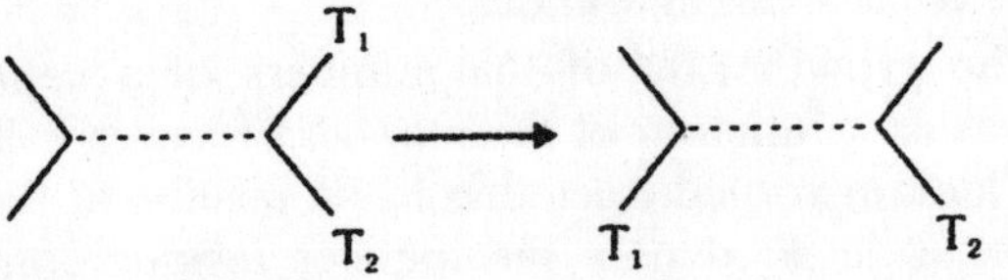

The number of acceptable indel configurations never decreases by such a rearrangement. This can be proven by showing that it applies for all symmetric (in the two ends) pairs of configurations for the connecting linear tree (proof left out because of space limitations). By doing a finite number of these changes, the linear tree will always be reached and it therefore has the maximal number of acceptable configurations.

For the linear rooted tree with n leaves, we can write the number of configurations of the different kinds as f_n, with $f_1 = [1\ 0\ 0]$:

$$f_n = [f_{n-1}Af_1^T \ f_{n-1}Bf_1^T \ f_{n-1}Cf_1^T] = f_{n-1}X = f_1X^{n-1}$$

$$\text{with } X = \begin{bmatrix} 6 & 2 & 1 \\ 3 & 2 & 1 \\ 1 & 1 & 1 \end{bmatrix}$$

By connecting the rooted tree with $n - 1$ leaves to the rooted tree with one leaf, the maximal number of acceptable configurations for the unrooted linear tree with n leaves (g_n) can be found by using (2)

$$g_n = f_{n-1}\begin{bmatrix}3 & 2 & 1\\ 2 & 1 & 0\\ 1 & 0 & 0\end{bmatrix} f_1^T = [1 \quad 0 \quad 0] X^{n-2}\begin{bmatrix}3\\ 2\\ 1\end{bmatrix}.$$

[3 2 1] is a linear combination of all three eigenvectors of X, so g_n grows exponentially as $O(m^n)$, where m is the maximal eigenvalue of X:

$$m = 3 + \frac{1}{\sqrt[3]{2}}\left(\sqrt[3]{21 + i\sqrt{59}} + \sqrt[3]{21 - i\sqrt{59}}\right) = 7.441622...$$

It follows that the memory complexity of the algorithm is $O(m^N L^{N-1})$. For specific numbers of acceptable indel configurations.

10.2.3.2 Time complexity

Since each acceptable residue configuration is applied to each acceptable indel configuration, the time complexity of the algorithm is at most proportional to the memory complexity times the number of acceptable residue configurations.

Here, the growth rate of the numbers of acceptable indel configurations as a function of the number of leaves will be found. First, let us look at rooted trees that has a residue at the root. For a rooted tree, a let x_a denote the number residue configurations with a residue at the root and where the vertices labeled with a residue form a connected graph. These configurations are acceptable, except for the possibility that no leaf has a residue.

Combining two rooted trees, a and b, to a new root vertex gives:

$$X_{ab} = (x_a + 1)\ (x_b + 1),$$

where the addition of one comes from having no residues in the corresponding subtree. Define

$$y = \log\ (x + 1)$$

Let x_i and y_i be the maximal x and y values, respectively, over the rooted tree topologies with i leaves. Also let $w_1 = 1$, and:

$$w_{2^{k+1}} = (w_{2^k} + 1)^2 \text{ and } z_{2^k} = \log(w_{2^k} + 1) \text{ for } k \geq 0$$

By recursively forming balanced trees, it is clear that $x_i \geq w_i$ for $i = 2^k$. It follows that the growth rat of x_i is at least equal to the growth rate of w_i (when defined, i.e. $i = 2^k$). To see that the growth rate is not higher, we will show that:

$$y_i \le \frac{2^{k+1}-i}{2^k} z_{2^k} + \frac{i-2^k}{2^k} z_{2^{k+1}} \text{ for } 2^k \le i < 2^{k+1}. \qquad \text{...(3)}$$

The right side is an affine function going through z_{2^k} for $i = 2^k$ and $z_{2^{k+1}}$ for $i = 2^{k+1}$. By showing this, we have strict limit for the growth of x_n between the values where w_n is defined. We have:

$$\frac{w_{2^{k+1}}+1}{w_{2^k}+1} < \frac{(w_{2^k}+1)^2}{(w_{2^{k-1}}+1)^2} \Rightarrow \frac{z_{2^{k+1}} - z_{2^k}}{2^{k+1}-2^k} < \frac{z_{2^k} - z_{2^{k-1}}}{2^k - 2^{k-1}}.$$

This means that the slope of the affine function in (3) decreases as k increases If (3) holds, we then also have :

$$y_i \le \frac{2^{k+1}-i}{2^k} z_{2^k} + \frac{i-2^k}{2^k} z_{2^{k+1}} \text{ for all k} \ge 0. \qquad \text{...(4)}$$

Equation (3) holds for $i = 1$ since $y_1 = z_1 = \log(2)$. Assume that (3) also holds for $i \le n$ and choose k so that $2^k \le n+1 < 2^{k+1}$. By using (4) we have:

$$y_i \le \frac{2^k - i}{2^{k-1}} z_{2^{k-1}} + \frac{i-2^{k-1}}{2^{k-1}} z_{2^k} \text{ for } i \le n.$$

It follows that

$$y_{n+1} \le \max_{1 \le i \le n} \log(\exp(y_i + y_{n+1-i}) + 1)$$

$$= \log\left(\exp\left(\frac{2^{k+1}-(n+1)}{2^{k-1}} z_{2^{k-1}} + \frac{(n+1)-2^k}{2^{k-1}} z_{2^k}\right) + 1\right) \qquad \text{...(5)}$$

$$\le \frac{2^{k+1}-(n+1)}{2^k} z_{2^k} + \frac{(n+1)-2^k}{2^k} z_{2^{k+1}}. \qquad \text{...(6)}$$

The fact that (5) is less than or equal to (6) for $2^k \le n+1 < 2^{k+1}$ can be seen by realizing that A) the two expressions are identical for $n + 1 = 2^k$ and for $n + 1 = 2^{k+1}$, and B) (6) is affine n, while (5) is convex in n, since $d^2/dx^2 \log(\exp(ax + b) + 1) > 0$ for $a > 0$. Induction leads us to conclude that (3) holds for all i. Aho and Sloane have shown that the recursion $q_{i+1} = q_i^2 + 1$ with $q_0 = 1$ grows as $O(a^{2i})$ with a given by:

$$a = \exp\left(\frac{1}{2}\log 2 + \frac{1}{4}\log\frac{5}{4} + \frac{1}{8}\log\frac{26}{25} + \frac{1}{16}\log\frac{677}{676} + \ldots\right)$$

$$= 1.502837 \ldots$$

Because of the affine relationship described in (3) and because $w_2k = q_{k-1}$, we see that x_n grows as $O(a^{2n})$, where $a^2 = 2.258518$.

These calculations were for rooted trees with a residue at the root and the possibility of no residue at any at any leaves. Due to the exponential growth rate of x_n, it can easily be shown that the contributions from the configurations with no residues at the root, and from those with no residues at any leaves, are negligible. Furthermore, the growth rate for unrooted trees is the same as for rooted trees. This means that the time complexity of the algorithms is $O(t^N L^N)$, with $t = ma^2 = 16.807040$. A slightly smaller exponential value that t is likely to apply for two reasons. (1) For a given indel configuration, only the residue configurations leading to acceptable indel configurations need to considered. (2) The worst case was the linear tree for the residue configurations and the balanced tree topology that is worst case for both types of configurations.

Much like for the indel configurations, specific numbers of acceptable residue configurations can be found by considering vectors representing different cases for rooted trees and then joining two rooted trees to form an unrooted one.

10.2.4 Results

The HOMSTRAD database consists of 1033 protein alignments (April 2003) which were made using structural information. These alignments will be considered correct in the following analyses. Three data sets of triple alignments were made by choosing sequences with pairwise identities in the ranges [15%; 25%], [25%; 35%], and [235%; 50%]. Since the starting and ending points of the sequences varied, indels in the ends of the alignments were removed.

As many triples as possible (16 to 18) within the identify ranges and with an average sequence length between 100 and 150 amino acids (after removing the ends) were randomly chosen from different HOMSTRAD families to form the data sets. For each triple alignment in the data sets, two analyses were made (1) All three possible pairwise alignments were performed and the percentage of correct matches was found. (2) The triple alignment was performed and the percentage of pairwise correct matches was found. For each data set, the gap introduction and extension costs were chosen to maximize the percentage of correct matches. Triple alignments are better than pairwise alignments for the less similar sequences. For the most similar sequences the performance is very close to identical.

Alignments made by the program Clustal W are a little better than with the algorithm presented here. This is not surprising given the more sophisticated position specific gap penalties of Clustal W and the simple distance matrix used here.

10.2.5 Conclusion

It has been shown how an optimal multiple parsimony alignment with affine gap cost over a tree can be found. The algorithm has a memory complexity of $O(7.442^N L^{N-1})$ and time complexity of $O(16.81^N L^N)$. This should be compared to the memory complexity of $O(L^{N-1})$ and time complexity of $O(2^N L^N)$ of the linear gap cost algorithm of combined with the Hirschberg approach. Triple alignment has been shown to outperform pairwise alignment on protein sequences.

It is likely that this trend will continue for more sequences, but time and memory become limiting factors in the calculations. However, it may be worthwhile to spend the extra computation time to do triple rather than use pairwise alignments in progressive alignment methods. Given a multiple alignment, the present method may be used to find its cost by limiting the dynamic programming to the prefixes defined by the alignment.

The time and memory complexity for this would be quite small. This may be useful in heuristic alignment methods that seek to improve alignments by local iterative adjustments. The advances made in this work may form the basis of more probabilistic methods. Statistical alignment methods may be developed that can utilize some of the same approaches for keeping track of gap introductions and extensions. This could also be useful in the context of analyses of multiple alignments using stochastic grammars, e.g., in RNA and protein structure prediction, and gene finding.

10.3 LARGE VERSION OF THE SMALL PARSIMONY PROBLEM

The relationship of biological sequences such as proteins and DNA can be described by a rooted tree, with a series of bifurcations corresponding to duplication in the past. Because of fluctuations in rates of evolution of such molecules it is not possible to determine the position of the root and the relative order of ancient duplications. When a rooted bifurcating tree is stripped of this information it becomes an un-rooted tree, where all internal vertices have three incident edges. Protein and DNA sequencing techniques allows the determination of the sequences at the leaves of such a tree.

The sequence that represent ancestral molecules are unobservable and must be reconstructed according to some principle. Usually the principle of parsimony is invoked: given a function that measures the distance between two sequences, choose ancestral sequences that minimize the total distance between two sequences, choose ancestral sequences that minimize the total distance between neighbors in the tree, i.e. the overall amount of evolution. The distance function defines the penalty, or cost of basic events such as substitutions and insertions/deletions (indels), and thus the distance between two sequences s_1 and s_2 is the "cheapest" process of evolution in terms of substitution and indels which transforms s_1 into s_2.

Application of the parsimony principle falls in two categories, depending on whether the phylogeny is known or not. If the phylogeny is known, it is the small parsimony problem; otherwise it is the large parsimony problem. This article will only address the small parsimony problem, but since it is hard version involving complete sequences, it is called a large version of the small parsimony problem. The simplest case where only substitutions have occurred and they all have cost 1 was first addressed by Fitch and independently by Hartigan. This was generalized to allow substitutions with arbitrary costs by Sankoff. The computing time used by these algorithms was of the order $O(n \,.\, k)$, where n is the length of the sequence and k the number of sequences.

The situation with two sequences subject to substitutions and indels of length 1 was solved by Sellers and Sankoff. This is called the *approximate string matching problem* and used $O(n^2)$ computing time. Sankoff combined the string matching and Fitch-Hartigan-Sankoff algorithms to devise an algorithm that reconstructs the history of a set of sequences if their phylogenetic tree is known. Its running time was $O(n^k \,.\, 2^k \,.\, k)$, which is prohibitive. Waterman et al. introduced sequence comparisons of two sequences with indels longer than one. Their algorithm uses $O(n^3)$ computing time. Gotoh gave an algorithm that uses time $O(n^2)$ if the gap penalty functions is of the form $g(l) = \alpha + \beta \cdot l$, where l is the length of the gap. Fredman generalized Gotoh's algorithm to three sequences if the gap penalty function was of the form $g(l) = \alpha$, i.e. the cost of an indel is independent of its length. Wang et al. studied the tree *alignment* problem where a multiple alignment is found by reconstructing the sequences at the internal nodes of a phylogeny minimizing the sum of all pairwise alignments between neighbours in the tree. In their approach gap costs are linear and the overall goal is to find a multiple alignment of the given sequences.

The algorithm was improved in where a polynomium time approximation scheme is given which yields a solution with a cost of at most 1.583 times the optimum, for sequences of length up to 200. Finally, much work exists on finding the best multiple alignment of k sequences with no phylogeny given, e.g. Hein devised an algorithm which finds an approximation to the optimal multiple alignment, of k sequences and simultaneously reconstructs their ancestors, given the tree. However, multiple alignment is not our aim in this article: we give an algorithm which *optimally* reconstructs the sequences at the internal nodes of a phylogeny, given an alignment of sequences at the leaves. The algorithm allows indels of any length and any sub additive gap penalty function (e.g. an affine function on the form $g(l) = \alpha + \beta \cdot l$ with $\alpha, \beta \geq 0$). Its worst-case time complexity is exponential, but in practice it is very fast and has been successfully tested for nine HIV sequences of length 9868.

10.3.1 Indels of Length Greater than One

Given a multiple alignment the problem arises of how to interpret gap symbols as indels. When only indels of length one are allowed each column can be treated separately. When longer indels are allowed the problem becomes non-trivial. There are gap symbols in sequences s_2 and s_3. The gaps partly overlap and they can be explained in two ways:

1. Two long indels have occurred: one on the evolutionary path between a and s_2 causing s_2's gaps, and one on the path between a and s_3 causing s_3's gaps. Consequently, a, the closest common relative of s_2 and s_3, has no gaps.
2. Three short indels have occurred: one on the path between s_1 and a causing gaps in the middle interval in both a, s_2 and s_3 in one go; one between a and s_2 accounting for s_2's gap in the firs column, and one between a and s_3 accounting for s_3's gap in the last column. With this explanation, a has a gap in the middle.

If we solve the problem the same way as when indels have length one, we would place the gaps as high (i.e., as far from the leaves) as possible in the tree, and so we would choose explanation 2—but that means that three indels have occurred whereas choosing explanation 1 means only two. On the other hand, since how both the *number* and *length* of the indels may vary between possible explanations, it is no longer obvious that the most parsimonious explanation is simply the one with the fewest indels. Whether to long indels explanation is simply the one with the fewest indels.

Whether two long indels are cheaper than three short ones depends on the chosen gap penalty function. With the function $g(l) = \alpha + \beta l$, where l is the length of the gap, the costs of the two explanations are $2(\alpha + 2\beta)$ versus $3(\alpha + \beta)$, respectively. Not only is it non-trivial to exhaustively list all possible explanations: it may also be non-trivial to find the cost of a given explanation. Consider the example, the gaps are caused by three indels; one of length 2 on the edge between s_2 and *a* causing s_2's gaps, one of length 1 on the edge between s_3 and *b* causing s_3's gaps, and one of length 3 on the edge between s_4 and *b* causing s_4's gaps. In explanation (2), a single indel occurring between *a* and *b* causes gaps in both s_3 and s_4 (and *b*) in the second column of the alignment. In explanation (3) this indel is pushed further back to the edge between s_1 and *a* now causing gaps in s_2, s_3 and s_4 all at once.

Depending on the direction of evolution, this indel is either an insertion happening *after*, or *a* deletion happening *before* the emergence of *b*, *s*3 and s_4. In both cases, the first and third alignment columns are in fact consecutive in *b*, s_3, and s_4. Therefore, though it might seem we need two indels to account for the two remaining gaps in s_4, they may be the result of only one. Thus, in this particular case, one indel can explain gaps in two non-adjacent alignment columns because the intermediate symbols have been deleted earlier or inserted later in the evolution. Concluding the example, we end up with the following costs for the three explanations: (1) $3\alpha + 6\beta$, (2) $3\alpha + 5\beta$, and (3) $3\alpha + 4\beta$. In passing, we observe that (3) is the optimal explanation for any (positive) choices of α and β. These examples hopefully illustrate the complexity of handling indels of length more than 1. In general two questions arise: "What are the consistent explanations of a given multiple alignment in terms of indels?" and second: "Given some gap penalty function, which explanation is optimal?" In the following we give an algorithm that solves the problem.

10.3.2 Algorithm

We assume we are given a multiple alignment M of $k > 2$ sequences S_1 ... S_k, and a phylogeny in the form of a binary evolutionary tree T relating these sequences. Thus each S_i is a leaf in T, and all other nodes in T have three neighbours. We do not take branch lengths into account. The length of M is n with the columns numbered 1 to n.

Each sequence is a string over some alphabet Σ extended with a gap symbol '–'. Further we are given a gap penalty function $g(l)$ defining the cost of a gap of length l. We assume g is subadditive: $g(l_1 + l_2) \leq g(l_1) + g(l_2)$. In the following, for brevity we will let Σ denote the set of nucleotides $\{A, C, G, T\}$, and we will use the affine gap cost function $g(l) = \alpha + \beta l$, although any alphabet applies as well as, e.g., logarithmic or root gap cost functions. We will write $[a : b]$ to denote the sequence of integers from a to b, including the endpoints a and b.

No sequence has all gaps, and neither does any column. We assume that aligned nucleotides are related (call it the R assumption): if sequences s_i and s_j both have a nucleotide in some column, the closest common ancestor of s_i and s_j also has a nucleotide in this position. It follows that *all* sequences on the path from s_i to s_j in T have nucleotides in this position. We make no assumption about the direction of time in the tree T. In other words, T is not *rooted*. For a leaf, we can talk about the unique *closest relative* since it has only one edge; for an internal node, the closest relative is not well-defined since internal nodes have three edges. If an edge connecting two nodes a and b is removed from T, the result is two disjoint rooted subtrees T_a and T_b rooted at a and b, respectively.

An insertion that occurs on this edge inserts nucleotides in all nodes in T_a and gaps in all nodes in T_b or vice versa. Similarly, a deletion converts nucleotides to gaps in all nodes in T_a and leaves the nucleotides in all nodes in T_b, or vice versa. In other words, *an indel affects a whole subtree* of T. Thus we can and will identify an indel by the subtree in which it inserts gaps. Note that the closest relative of a subtree $T' \subset T$ is well-defined: it is unique node $n \notin T$ which has an edge to the root of T'. It is easily shown that a tree with $k > 2$ leaves has $4k - 5$ different non-empty subtrees. We can represent a subtree T' of T unambiguously by its leaves; of course not all leaf subsets represent subtrees. This tree with $k = 4$ leaves has $4k - 5 = 11$ different subtrees (counting T itself); e.g., $\{s_1\}$, $\{s_3, s_4\}$, and $\{s_1, s_2, s_4\}$. Note that for example, $\{s_2, s_3\}$ is *not* a subtree: one can not cut one edge and end up with a subtree containing exactly the leaves $\{s_2, s_3\}$.

10.3.2.1 Construction of a gap graph

An alignment column has a *gap configuration* given by the sequences that have gap symbols in this column. A maximal interval $I \subseteq [1 : n]$ where alle columns have the same gap configuration is called a *gap interval*. E.g., the alignment has four gap intervals: $I_1 = [1 : 1]$, $I_2 = [2 : 2]$, $I_3 = [3 : 4]$, and $I_4 = [5 : 7]$. Gap intervals are important since the indels used to explain the gaps in one column of a gap interval can be extended to explain the remaining columns without the cost of introducing extra indels (by subadditivity of the gap cost function this is cheaper, so indels are never broken off in the middle of a gap interval). Recall that each leaf in the evolutionary tree T corresponds to a sequence. The *tree covering* of a gap interval I is the minimal forest F of subtrees of T where

(C1) All leaves of T with gap symbols in I belong to a subtree in F.

(C2) All leaves in each subtree in F have gap symbols in I

Thus, a tree covering of I "covers" exactly the sequences that have gaps in I and not the ones that have nucleotides. Note that a tree covering is minimal in the *number* of subtrees that satisfy the above conditions. In other words, each subtree is maximal: it cannot be extended without including leaves with nucleotides in I.

In the following, we will represent a gap interval by a pair (F, I), where I is the gap interval itself and F is its associated tree covering. Look again, in I_1, to satisfy conditions $C1$ we need to cover the sequence s_3 only. Therefore, the tree covering F_1 associated with I_1 is $\{\{s_3\}\}$, a forest of only one subtree. I_2 and I_3 are also covered by single subtree forests: $F_2 = \{\{s_3, s_4\}\}$, and $F_3 = \{\{s_1, s_2, s_3\}\}$. Finally, to cover I_4 we need two subtrees: $F_4 = \{\{s_1\}\}, \{\{s_4\}\}$. This is a minimal forest: any subtree with more than one leaf would include s_2 or s_3 which have nucleotides in I_4, and thus the subtree would violate conditions C2.

For now we focus on the gaps, and so we simply represent the alignment by its constituent gap intervals, i.e. the set $\{\{F_i, I_k\}\}$, where, for all i, interval I_i precedes interval I_{i+1} in the alignment and $\cup I_i$ is a disjoint partition of $[1 : n]$. We call such a set a *gap division*. A gap division D now induces a *gap graph* in the following manner. Each gap interval $(F_i, I_i) \in D$ gives rise to a set of vertices in the graph; for each subtree T' in the tree covering F_i, a vertex $(T')I_i$ is created. We say that the vertex (T') lies in the interval I_i.

We observe that since no alignment columns have all gaps, the subtree T' of any vertex is a true subtree of the whole evolutionary tree T and so T' has a closest relative.

As we shall see, a vertex $(T')I_i$ represents a *potential* indel: an indel occurring on the edge between the root of T' and its closest relative could be the cause of all the gaps in the sequences in T' in the gap interval I_i. Consider a gap graph with two vertices $(T_1)I_1$ and $(T_2)I_2$ that lie in consecutive intervals I_1 and I_2. Then the subtree T_1 has one of five possible relationships with T_2:

- *(twin)* $T_1 = T_2$
- *(son)* $T_1 \subset T_2$
- *(father)* $T_1 \supset T_2$
- *(cousin)* $T_1 \cap T_2 \neq \emptyset$, but $T_1 \not\subseteq T_2$ and $T_1 \not\supseteq T_2$
- *(unrelated)* $T_1 \cap T_2 = \emptyset$

These five relationships are manifested in the gap graph as follows:

1. If T_1 and T_2 are twins the two vertices are *merged* into a vertex $(T_1)_{I_1 \cup I_2}$ that lies in the interval $I_1 \cup I_2$ (note that this new interval is not a gap interval since it does not have the same gap configuration everywhere).
2. If T_1 is the son of T_2 a *directed edge* is created form the vertex $(T_1)_{I_1}$ to the vertex $(T_2)_{I_2}$.
3. If T_1 is the father of T_2 is *directed edge* is created form the vertex $(T_2)_{I_2}$ to the vertex $(T_1)_{I_1}$.
4. If T_1 and T_2 are cousins, an *undirected zigzag edge* is created between the vertices $(T_1)_{I_1}$ and $(T_2)_{I_2}$.
5. If T_1 and T_2 are unrelated, no edges are created.

Thus to construct a gap graph from a gap division $D = \{(F_i, I_i)\}$ we first create a vertex for each subtree in each tree covering F_i. Then we traverse the gap intervals I_i looking at two consecutive intervals at a time and merge any twin vertices, create directed edges between fathers and sons, and create undirected zigzag edges between cousins. Note that twin vertices may be extended to more than two gap intervals. We next proceed to a full example. It has four gap intervals $I_1, ...I_4$, each with its own tree covering $F_1, ...F_4$. Each subtree in each tree covering induces a vertex in the gap graph, so initially we get the six vertices $\{2\}_{I_1}$, $\{2\}_{I_2}$, $\{4, 5\}_{I_2}$, $\{1, 2, 3, 4\}_{I_3}$, $\{1, 2\}_{I_4}$, and $\{4\}_{I_4}$. Traversing the gap intervals, first we merge the two $\{2\}$-vertices since they lie in consecutive intervals, and then we create a directed edge from this new vertex to $\{1,2,3,4\}_{I_3}$ since $\{2\} \subset \{1, 2, 3, 4\}$. Next we create an undirected zigzag edge between $\{4,5\}_{I_2}$ and $\{1,2,3,4\}_{I_3}$ since neither is contained in the other while they still share the leaf 4.

Finally we create directed edges going form $\{1,2\}_{I_4}$ and $\{4\}_{I_4}$ to $\{1, 2, 3, 4\}_{I_3}$. For clarity we write the subtree of a gap graph vertex as either a variable in parentheses, like in $(T')_{I'}$, or a specific list of leaves with no parentheses, like in $\{1,2\}_{I'}$. Our algorithm now works with the gap graph to find the most parsimonious set of indels that explains the gaps in the alignment, given a gap penalty function $g(l) = \alpha + \beta l$. We first go through a preprocessing phase in which we reduce the potentially very large and complex gap graph. In the second phase, we resolve the reduced graph to find the optimal solution.

10.3.2.2 Preprocessing the gap graph

As already said, indel $\theta_{I,T}$ results in gaps in all nodes of subtree T in some (set of) interval(s) I of the alignment. The following theorem establishes a strong connection between indels and gap graph vertices. Before we can prove the theorem, we need a small lemma.

Lemma. Given a subtree T' and a sequences S of gap graph vertices $(T_k)_{Ik}$ such that $\forall k$, (1) $T' \subset T_k$, and (2) $T_k \subseteq T_{k+1}$ or $T_k \supseteq T_{k+1}$ (i.e., no neighbors are cousins). Then there exists a subtree $T \supset T'$ such that $T \subseteq T_k$ for all k.

Proof. Since T' is a real subtree of the T_k's, T' has a closest relative a, and each T_k consists of at least $T \cup a$. Now a cannot be a leaf since the T_k's are real subtrees of the whole evolutionary tree (if some $T_{'''}$ were the whole tree, all sequences would have gaps in the associated interval $I_{'''}$). Thus, a has relatives b and c with associated (possibly empty) subtrees T_b and T_c, and the evolutionary tree looks like the figure below. Consider T_1, the first subtree in the sequence S. Since it contains T' and a, it must also contain either T_b or T_c, but not both. Assume without loss of generality that it contains T_b. Then its neighbour in the sequence, T_2 contains T_b as well. If not T_1 and T_2 would be cousins: they have T' in common, T_1 contains T_b (and not all of T_c), but T_2 would contain T_c. This argument is easily extended by induction to show that all subtrees T_k contains T_b. Let $T = T' \cup a \cup T_b$. Then $T \supset T'$ and $T \subseteq T_k$ for all $T_k \in S$.

Theorem. Let E be an optimal set of indels, explaining an alignment M with respect to a gap cost function $g(l) = \alpha + \beta l$, and let G be the gap graph associated with M. Let $\theta_{I,T} \in E$ be an indel that creates gaps in the nodes of the subtree T in the interval I. Then,

1. $\exists$ a vertex $(T')_{I'}$ in G with $I' \subseteq I$ such that $T' = T$, *or*
2. $\exists$ two cousins in I, $(T_1)_{I1}$ and $(T_2)_{I2}$ such that $T \subseteq T_1$ and $T \subseteq T_2$.

Thus either the indel corresponds to (The subtree of) an existing vertex that lies somewhere in the indels' interval I, or it "crosses" a cousin edge in I.

Proof. Let $I = \cup I_k$ where the I_k are the disjoint gap intervals in the gap graph G. All leaves of T have gaps in all of I, and so these leaves are covered in each gap interval I_k and G. Thus, in each I_k there exists a vertex $(T_k)_{Ik}$ such that $T \subseteq T_k$. Let S be the set of subtrees of these vertices. If two neighbour subtrees in S, T_k and T_{k+1}, are cousins, we are done. Otherwise, we have that $T_k \subseteq T_{k+1}$ or $T_k \supseteq T_{k+1}$ for all k. Assume that $\forall T_k : T \subset T_k$.

Lemma ensures the existence of subtree T^* such that $T \subset T^* \subseteq T_k$ for all $T_k \in S$. Recalling that each T_k comes from a vertex $(T_k)_{Ik}$ in the gap graph, all nodes/sequences in T^* must have gaps in I. By assumption the optimal explanation E contains one indel $\theta_{I,T}$ that accounts for the gaps in T interval I. The cost of this indel is $\alpha + \beta l$, if we let l be the length of I. Further, some number of indels (say $m > 0$) must account for the gaps in the nodes in $T^* \backslash T$ in interval I. Note that these m indels may stretch beyond I, so their total cost is $m\alpha + \beta(ml + d)$, where we sum the extra lengths beyond I in d. Thus, we can write the total cost g of the optimal explanation E as

$$g = (\alpha + \beta l) + m\alpha + \beta(ml + d) + q$$

where q is the cost of all indels accounting for other gaps. Consider the explanation E^* obtained from E by replacing $\theta_{I,T}$ with the indel θ_{I,T^*} which creates gaps in interval I in all nodes in the subtree T^*. The cost of this new indel is still $\alpha + \beta l$, while we no longer have to let the m extended indels affect interval I. Thus we save βml on the cost, and the total cost g^* of the explanation E^* is

$$g^* = (\alpha + \beta l) + m\alpha + \beta d + q < g$$

But this contradicts the optimality of E. Therefore, T must be equal to at least one of the T_k's, and so the gap graph G must have a vertex $(T_k)_{Ik}$ that corresponds to the indel $\theta_{I,T}$.

In general, given a vertex $(T')_I$, the question is whether it corresponds directly to an ideal happening on the edge between, the root of T' and its closest relative, or whether the gaps of T' resulted from several indels happening inside T'. In the latter case we say that the vertex is *decomposed*, otherwise it is *confirmed*. Theorem tells us where to look when searching for the indels in optimal explanation: Candidates are found among the (Subtrees of the) gap graph vertices and the *intersection between cousins*. In a (part of a) gap graph with no cousin edges all indels in the optimal explanation corresponds directly to subtrees of existing vertices.

If a cousin edge is present, however, the optimal explanation may contain indels not represented directly as vertices; such as indel must then correspond to a subtree which is contained in both the cousins' subtrees and thus lies in their intersection. Thus theorem serves two causes. First it pins out cousin vertices as special cases, and second it justifies the perhaps intuitively reasonable notion that one should not place indels "lower than necessary in the tree: if a vertex $\{1, 2\}_I$ is not connected to leaf vertices $\{1\}_{Ij}$ and $\{2\}_{Ik}$ anywhere (and no cousins either), the corresponding graphs in sequences s_1 and s_2 do not both extend beyond I, and so there is no reason not to make the gaps in s_1 and s_2 in I in one go, i.e. on the edge between the root of the subtree {1, 2} and its closest relative. In fact we can say that if a vertex is decomposed, all of the decomposing indels continue and help explain other vertices in the gap graph. We formalize this in a lemma (which follows easily from Theorem so we omit the proof) and elaborate in it a theorem.

Lemma. If a gap graph vertex $(T)_I$ is decomposed in the optimal explanation of the corresponding alignment, then all the indels explaining gaps in T extend beyond interval I.

Theorem. If a gap graph vertex $(T')_I$ is decomposed in the optimal explanation of the corresponding alignment, the decomposing indels do not all extend form I in the same direction.

Proof. Let $(T')_I$ be decomposed into m indels. By Lemma, all decomposing indels extend beyond interval I. Assume they all extend in the same direction.

Let r be the root of T' and let a be the closest relative of T'. Let l be the length of I. Let $g(l) = \alpha + \beta l$ be the cost of an indel of length l, and let g^* be the cost of the optimal explanation E^* of all gaps in the alignment, (i.e. the explanation that minimizes the total cost over all indels). In this explanation, the gaps in the leaves of T' are explained by the m indels that extend beyond I. The total cost of these indels can be written as $m\alpha + \beta(ml + d)$ (Each such indel has some length greater than l and we sum these extra lengths in the constant d). We can then write

$$g^* = m\alpha + \beta(ml + d)\ q$$

where q is the total cost of all other indels. Since T' decomposed into at least two indels, $m \geq 2$.

These m indels all start in interval I and by assumption they extend in the same direction. If they all end in exactly the same column c_e in the alignment, we replace them with only one indel occurring between r and a and extending from I to c_e. Such an explanation would trivially be cheaper than g^*, contradicting the optimality of E^*. Thus, the m indels do not have equal lengths. Let c be the first alignment column where one of the m indels ends (call this indel θ_1). The column c is outside of I since θ_1 extends beyond I. Define I_x to be the extended interval that starts with the includes I and ends in column c; let its length be l_x.

Consider the explanation F_1 obtained from E^* by replacing θ_1, with the indel $\theta_{Ix,\,T'}$ occurring between r and a in the evolutionary tree and extending all through the interval I_x. This new indels explains the gaps in all T' in all of I_x, including I, while $m - 1$, of the original indels (all except θ_1) still have to be made since they go beyond I_x; however, as we have covered all gaps in I_x with $\theta_{Ix.T'}$ we save $\beta m l_x$ on the cost. Therefore, the cost of explanation E_1 is

$$g_1 = (\alpha + \beta l_x) + (m - 1)\alpha + \beta(ml + d) + q - \beta m l_x =$$
$$m\alpha + \beta(ml + d) + q - \beta(ml_x - l_x) \quad = g^* - \beta l_x(m - 1)$$

Since $m \geq 2$ we have that $\beta l_x(m - 1) > 0$ and thus $g_1 < g^*$. But this contradicts the optimality of E^* and so our assumption must be wrong. Therefore, the decomposing indels do not all extend in the same direction.

Recall that a gap graph vertex represents gaps in a subtree of sequences in some region of an alignment. The vertex may have edges to other vertices, and these edges can have three different types (in-, out-, and cousin edges), It turns out that vertices with certain configurations of edges can immediately be confirmed to correspond to optimal indels. In other words certain types of vertices can be "decided" immediately and locally.

Using the two strong Theorems and we can characterize formally exactly which types of vertices we can immediately decide. We will not go into the details here but simply give the results. A vertex is called a *leaf vertex* if its subtree consists of one leaf node only. Such a vertex can be confirmed right away and thus corresponds to an optimal indel occurring between this node and its closest relative in the tree. Likewise, *orphans* and *end vertices* can be confirmed: they are vertices with no edges and vertices with edges on one side only, respectively.

All sequences in the subtree of an orphan have nucleotides on either side of its interval Consider a vertex (T')$_I$ where *some but not all* of the leaves of its subtree T' have gaps in an adjacent interval. Such a vertex is called a *patriarch*. A patriarch has at least one edge, but its subtree also has at least one leaf which is not included in a subtree in any vertex in an adjacent interval. We say that such a leaf has *solo gaps*: it has gaps in interval *I* but nucleotides on both sides. Patriarchs may also be confirmed immediately; intuitively, since it is necessary to spend an indel on the solo gaps, this indel may as well create all other gaps of the vertex' subtree also.

A patriarch has another important property: after its confirmation, its edges are removed and new edges are created directly between its former neighbours on either side. A patriarch has no out-edges, and its subtree contains the subtrees of its in-edge neighbours; thus, by the same argument explained in connection with the intervals associated with the patriarch's neighbours are in fact consecutive and so the neighbours may be connected directly if they share leaves.

A special case is when the patriarch has two cousins; in this situation the edges cannot be removed since we have to keep the information that no indel with a larger subtree than that of the patriarch can "pass" the patriarch and cause gaps on both of its sides in one go. A few other reduction rules help reduce the gap graph further. The preprocessing phase makes use of these theoretical results by going over the gap graph in a series of passes, each pass reducing the graph by local applications of the reduction rules. Since the removal of edges following a patriarch confirmation may turn previously undecidable vertices decidable, several passes are performed.

10.3.2.3 Resolving the reduced gap graph

Once the preprocessing is done, the gap graph has been reduced significantly. In other words, most the gaps have been explained. We now turn to the *chains* of the gap graph: two vertices belong to the same chain if an only if there exists a path connecting the two which does not cross a leaf vertex. Thus chains often end in leaf vertices, some of which may belong to two chains. The chains are important since they can be dealt with independently: indels causing gaps in the vertices of one chain could not have caused gaps in vertices of another chain. In the second phase of our algorithm, we analyze each chain in turn. Looking first at chain *a*, we see that only the vertex is still undecided (the others being leaf vertices).

Referring to Theorem, we not that since there are no cousin edges in this chain, the only indels we need to consider for the optimal explanation are the ones already represented by vertices in the chain. That means there are only two possible explanations for the gaps represented by the vertex (123): (1). Either they are the result of one indel occurring between the root of the subtree {1,2,3} and its closest relative (in which case we confirm the vertex), or (2) they are the result of three indels occurring between 2 and its closest relative, 1 and its closest relative, and 3 and its closest relative, respectively, extending the already confirmed neighboring indels/ vertices (in which case we decompose the vertex). If we let l_{123} be the length of its associated alignment interval, the extra cost for resolving the vertex (123) in each of these two explanations is $\alpha + \beta l$ versus $3\beta l$.

For the optimal explanation we simply choose the cheaper one (in general, if several options are optimal, we choose one arbitrarily). Chain *b* has three undecided vertices: (12), (1234), and (345). Moreover, the latter two are cousins and so by Theorem we need to consider indels whose subtrees lie in their intersection. That gives two additional potential indels not already represented in the chain by vertices, namely {3} and {4}. Interestingly the vertex (12) can therefore be confirmed right away: its only possible decomposition is in subtrees {1} and {2}, but since {2} is neither present as a vertex nor lies in the cousins intersection, this indel could never be part of an optimal explanation.

For the same reason, the vertex (1234) may either be confirmed or decomposed in one way only, namely with three indels with the subtrees {1, 2}, {3}, and {4}. And finally the vertex (345) may either be confirmed or decomposed in three indels with the subtrees {3}, {4}, and {5} (this time, Theorem dictates that the indel {4, 5} is not an option in chain (b). In total therefore we have four combinations which we need to check in this chain. Lastly, in chain *c* we may either confirm the vertex (45) or decompose it in {4} and {5}. Checking all combinations we find the optimal indels. Each causes gaps in some set of columns in a full subtree, and by the *R* assumption, we lace anonymous nucleotides in the internal nodes in the rest of the tree in the same columns. It remains to optimally name these nucleotides. For each column we ignore the subtrees that have all gaps and get a still connected tree with all nucleotides, and on this tree we now perform the Finch-Hartigan-Sankoff algorithm as a final step to optimally assign nucleotides to the internal nodes.

10.3.3 Performance and Future Work

Checking the cost of each combination of confirmation/ decomposition of all the vertices in a chain cannot be done in constant time. The combination in chain *b* of decomposing both (1234) and (345) we have to look at both decompositions to see that the decomposing indels {3} and {4} can extend to both vertices and do not have to be opened a new. Note also that since we have to check "all with all", the number of combinations to check is exponential in the number of undecided vertices in the chain. I.e., it is critical that as many edges as possible are removed in the preprocessing phase. If the tree is large, some vertices may have many possible decompositions; if the alignment has many "crisscrossing" gaps, the chains may become very long.

Encountering, e.g., a chain with 11 undecided vertices each with five possible decompositions (including not decomposing) except two which have 10 decompositions, we have in total $5^9 . 10^2 =$ 195312500, close to 200 million, combinations to check (each taking some hard to characterize but more than constant amount of time) for this chain alone. Thus, the worst case time complexity is exponential in the length of the alignment. On the other hand, if the gaps mostly fall in straight blocks delimited by full columns of nucleotides, chains will be short and the running time will be very fast. Thus, the running time of our algorithm is extremely dependent on the particular problem instance. To demonstrate this, we have tested the algorithm on a huge number of different alignments with some ratio of gaps scattered individually across all sequences. We wanted to challenge our algorithm with very had data, and a long such alignment with a high gap ratio of 60% is indeed a very hard problem to solve.

Real alignments do not look like this: firs they do not have that many gaps unless the aligned sequences are only very distantly related. And second, since gaps appear as the result of indels, they do appear in connected blocks and not as much as independent single gaps. For this reason we also did some experiments on alignments with only 40% and 50% gaps, and finally we ran the algorithm on an alignment of nine full HIV-1 subtype genomes.

We show the average performance for random alignments with 60% gaps over different alignment lengths with 1000 trials for each, on a tree with nine leaves. For an alignment of length 10000 the average time was around 40 seconds. For lengths below 6000, the time is less than 10 seconds. The curve does not appear smooth; this is due to the above mentioned dependency on the a particular problem instance resulting in a huge variance. Since it would have been to time consuming to actually find the optimal explanation in 1000 trials with no preprocessing, we instead report the number of combinations it would have been necessary to check.

Again the variance is huge so we report the 95% fractiles (i.e. a value of 1000 means that in 95% of the trials, the number of explanations to check was at most 10000). We did experiments with and without preprocessing, and to also get an idea of the performance for different tree sizes we used both a tree with nine leaves and one with six leaves. For example, for alignment length 10000 and a tree with nine leaves, the numbers are about 500.000 with preprocessing versus about 100 million without preprocessing: i.e. preprocessing reduces the number of combinations to check by a factor of about 2000 in this case. We also observe that for the smaller tree, the number of combinations to check is substantially smaller. This performance is contrasted by our results for a tree with nine leaves and alignments with only 40% or 50% gaps: for all lengths up to 8000, the average running time was less than one second.

We also analyzed a full genome alignment of B.ES.89.S61K15, B.FR.83.HXB2, B.GA.88.OYI, B.GB.83CAM1, B.NL. 86.3202A21, B.TW.94.TWCYS, B.US.86.AD87, B.US. 84.Ny5GC, and B.US.83.SF2, nine HIV-1 subtypes retrieved from the Los Alamos HIV database. After constructing a phylogeny using the neighbour-joining Sanger Institute Quicktree software we had a tree with nine leaves and an alignment of length 9868 (this example only serves to demonstrate our algorithm, the method chosen to obtain phylogeny and alignment were arbitrary). Solving this problem again took less than one second. Because of the great similarity of the sequences, the resulting gap graph was small and simple. We are interested in running our program on more, real data sets with alignments and trees with shorter alignments (100-500 leaves, length < 500) is an interesting application which we have not yet looked at. Also, we intend to combine the algorithm with some heuristic search method so as to *find* an approximately optimal alignment which fits the tree, rather than assume it is given (a variant of the tree alignment problem). Our might consider our algorithm a way of ranking a given combination of tree and alignment using a gap cost function and a substitution matrix, and so trying different alignments, or even trees, might actually improve the data.

10.4 GRAPH ALGORITHMS

Many bioinformatics algorithms may be formulated in the language of *graph theory*. The use of the word "graph" here is different than in many physical science contexts: we do not mean a chart of data in a Cartesian coordinate system. In order to work with graphs, we will need to define a few concepts that may not appear at first to be particularly well motivated by biological examples, but after introducing some of the mathematical theory we will show how powerful they can be in such bioinformatics applications as DNA sequencing and protein identification.

10.4.1 Graph

Figure 10.4 (a) shows two white and two black knights on 3 × 3 chessboard. Can they move using the usual chess knight's moves, to occupy the position shown in figure 10.4 (b)? Needless to say, two knights cannot occupy the same square while they are moving.

Figure 10.5b represents the chessboard as a set of nine points. Two points are connected by a line if moving from one point to another is a valid knight move. Figure 10.5c shows an equivalent representation of the resulting diagram that reveals that knights move around a "cycle" formed by points 1, 6, 7, 2, 9, 4, 3 and 8. Every knight's move on the chessboard corresponds to moving to a neighboring point in the diagram, in either a clockwise or counterclockwise direction. Therefore the white-white-black black knight arrangement cannot be transformed into the alternating white-black-white-black arrangement.

Diagrams with collections of points connected by lines are examples of *graphs*. The points are called *vertices* and lines are called *edges*. A simple graph shown in figure 10.6, consists of five

(a)

(b)

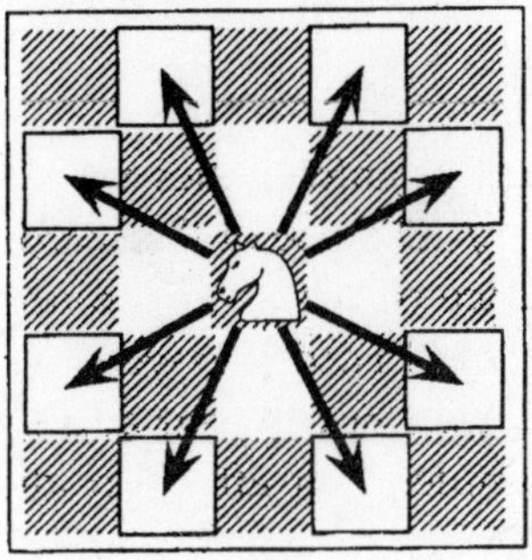

(c)

Figure 10.4. Two configuration of four knights on a chessboard. Can you use valid knight moves to turn the configuration in (a) into the configuration in (b)? Valid knight moves are shown in (c)

vertices and six edges. We denote a graph by $G = G(V, E)$ and describe it by its set of vertices V and set of edges E (every edge an be written as a pair of vertices). The graph in figure 10.6 is described by the vertex set $V = \{a, b, c, d, e\}$ and the edge set $E = \{(a, b), (a, c)\ (b, c), (b, d), (c, d), (c, e)\}$.

The way the graph is actually drawn in irrelevant; two graphs with the same vertex and edge sets are equivalent, even if the particular pictures that represent the graph appear different. The only important feature of a graph is which vertices are connected and which are not.

Figure 10.7 represents another chessboard obtained from a 4 × 4 chessboard by removing the four corner squares. Can a knight travel around this board, pass through each square exactly once, and return to the same square it started on? Figure 10.7 (b) shows a rather complex graph with twelve vertices and sixteen edges revealing all possible knight moves. However, rearranging the vertices reveals the cycle that describes the correct sequence of moves. The number of edges incident to a given vertex v is called the *degree* of

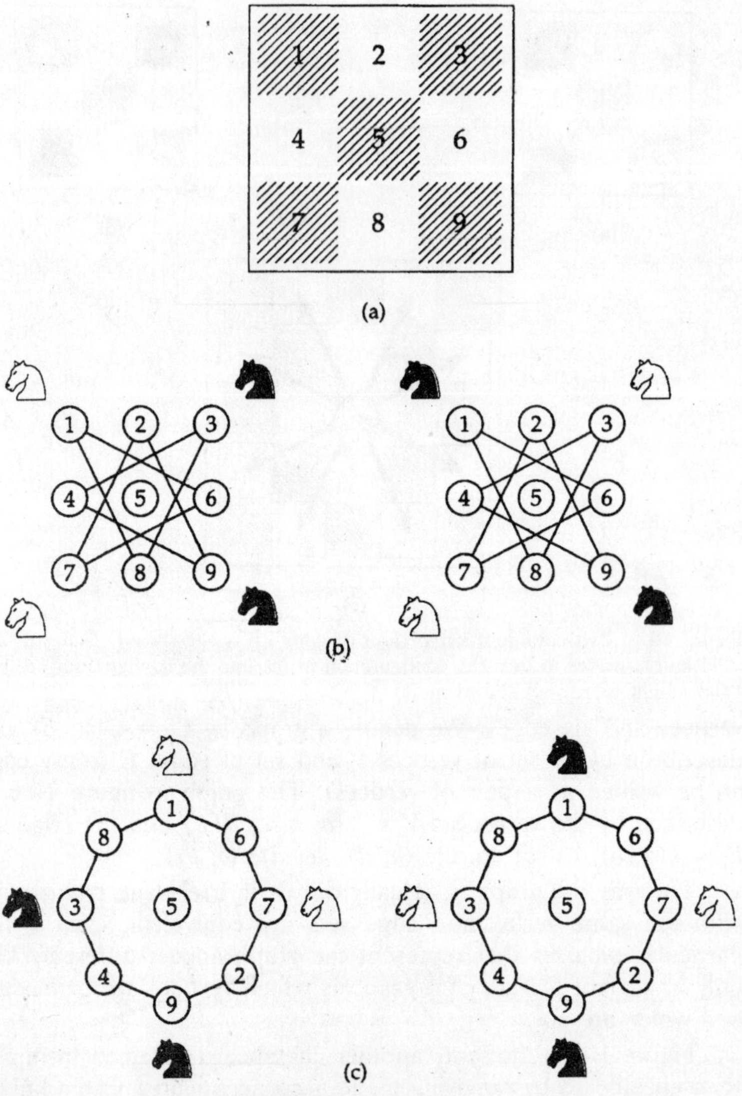

Figure 10.5. A graph representation of a chessboard. A knight sitting on some square can reach any of the squares attached to that square by an edge.

the vertex and is denoted $d(v)$. For example, vertex 2 in figure 10.7 (c) has degree 3 while vertex 4 has degree 2. The sum of degrees of all 12 vertices is, in this case, 32 (8 vertices of degree 3 and 4 vertices of degree 2), twice the number of edges in the graph. This

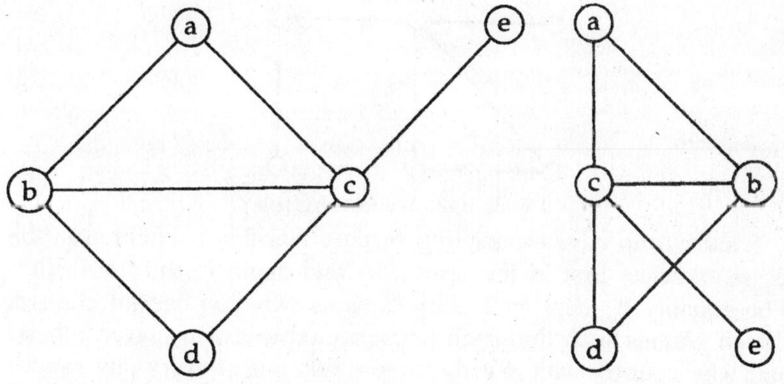

Figure 10.6. Two equivalent representations of a simple graph with five vertices and six edges.

is not a coincidence: for every graph G with vertex set V and edge set E, $\Sigma_{v \in V} d(v) = 2 \cdot |E|$.

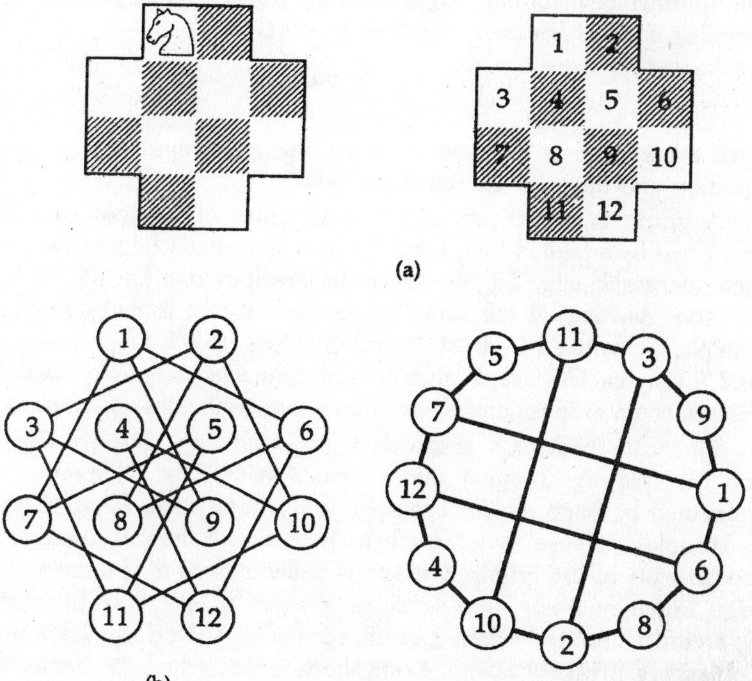

Figure 10.7. The knight's tour through the twelve square in part (a) can be seen by constructing a graph (b) and rearranging its vertices in a clear way (c).

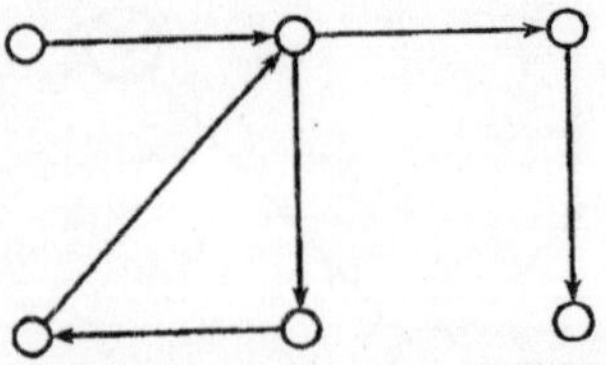

Figure 10.8. A directed graph.

Indeed, an edge connecting vertices v and w is counted in the $\Sigma_{v \in V} d(v)$ twice: first in the term $d(v)$ and again in the term $d(w)$. The equality $\Sigma_{v \in V} d(v) = 2 \cdot |E|$ explains why you cannot connect fifteen phones such that each is connected to exactly seven others, and why country with exactly three roads out of every city cannot have precisely 100 roads. Many bioinformatics problems make use of *directed* graphs, in which every edge is directed from one vertex to another, as shown by the arrows in figure 10.8.

Every vertex v in a directed graph is characterized by *indegree*(v) (the number of incoming edges) and *outdegree*(v) (the number of outgoing edges). For every directed graph $G(V, E)$,

$$\sum_{v \in V} indegree(v) = \sum_{v \in V} outdegree(v),$$

since every edge is counted once on the right-hand side of the equation and once on the left hand side.

A graph is called *connected* if all pairs of vertices can be connected by a *path*, which is continuous sequence of edges, where each successive edge begins where the previous one left off. Paths that start and end at the same vertex are referred to as *cycle*. For example, the paths (3-2-10-11-3) and (3-2-8-6-12-7-5-11-3) in figure 10.7 (c) are cycles. Graphs that are not connected are *disconnected*. Disconnected graphs can be partitioned into *connected components*.

One can think of a graph as a map showing cites (vertices) and the freeways (edges) that connect them. Not all cities are connected by freeways: for example, you cannot drive from Miami to Honolulu. These two cities belong to two different connected components of the graph. A graph is called *complete* if there is an edge between every two vertices. *Graph theory* was born in eighteenth century when Leonhard Euler solved the famous Konigsberg Bridge problem. Konigsberg is located on the banks of the Pregel River, with a small island in the middle. The various parts of the city are connected by bridges and Euler was interested in whether he could arrange a tour of the city in such a way that

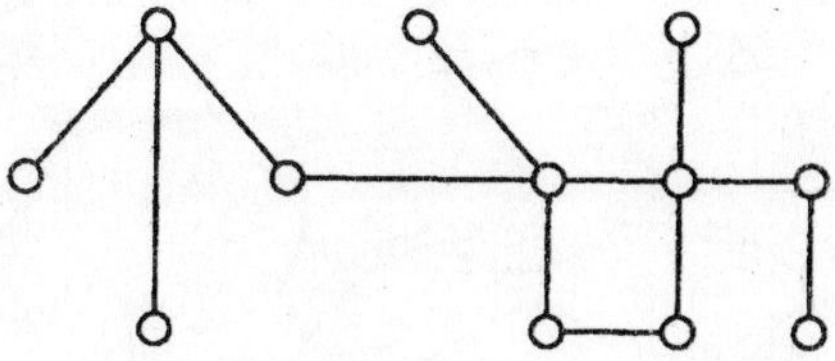

(a)

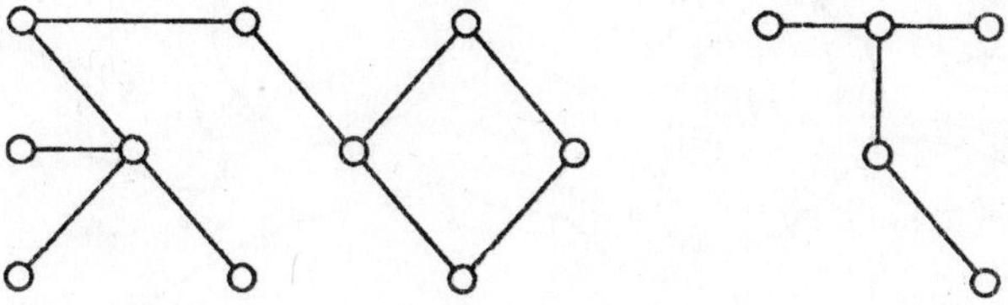

(b)

Figure 10.9. A connected (a) and a disconnected (b) graph.

the tour visits each bridge exactly once. For Konigsberg this turned out to be impossible, but Euler basically invented an algorithm to solve this problem for any city. Figure 10.11 shows a city map with ten islands and sixteen bridges, as well as the transformation of the map into a graph with ten vertices and sixteen edges (every island corresponds to a vertex and every bridge corresponds to an edge). After this transformation, the Bridge Obsession problem turns into the Eulerian Cycle problem that was solved by Euler and later found thousands of a applications in different areas of science and engineering: After the Konigsberg Bridge problem was solved, graph

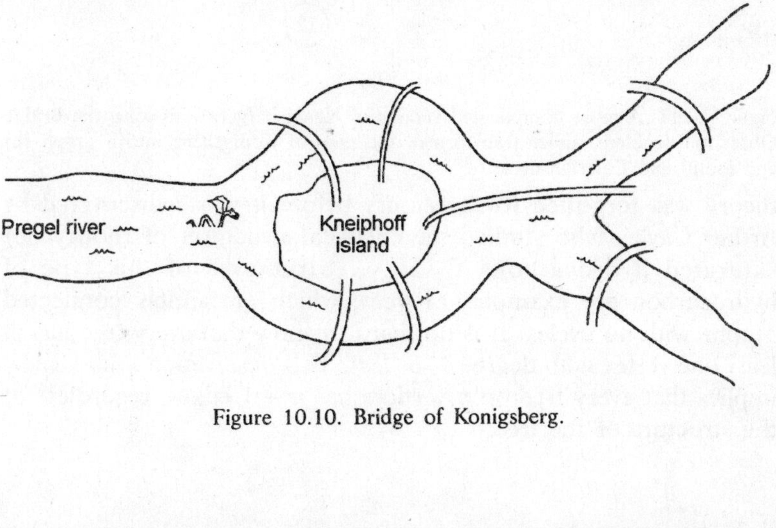

Figure 10.10. Bridge of Konigsberg.

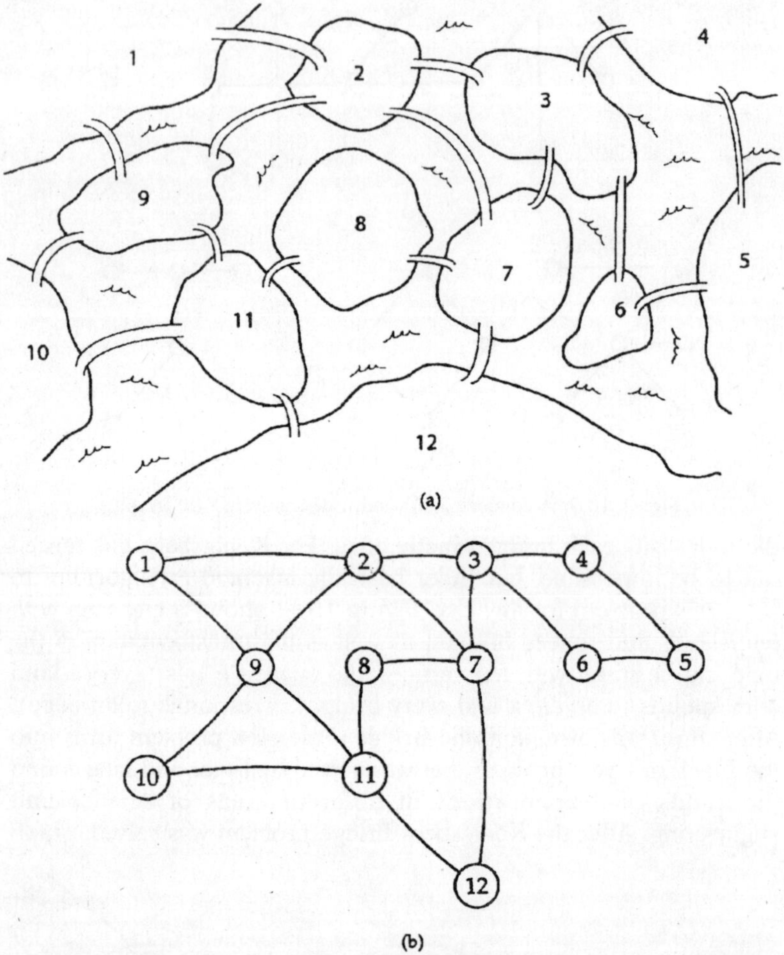

Figure 10.11. A more complicated version of Konigsberg (a). To solve the Bridge Obsession problem, Euler transformed the map of Konigsberg into a graph (b) and found and Eulerian cycle.

theory was forgotten for a century before it was rediscovered by Arthur Cayley who studied the chemical structures of (noncyclic) saturated hydrocarbons $C_n H_{2n+2}$. Structures of this type of hydrocarbon are examples of *trees,* which are simply connected graphs with no cycles. It is not hard to show that every tree has at least one vertex with degree 1, or *leaf*. This observation immediately implies that every tree on *n* vertices has $n - 1$ edges, regardless of the structure of the tree.

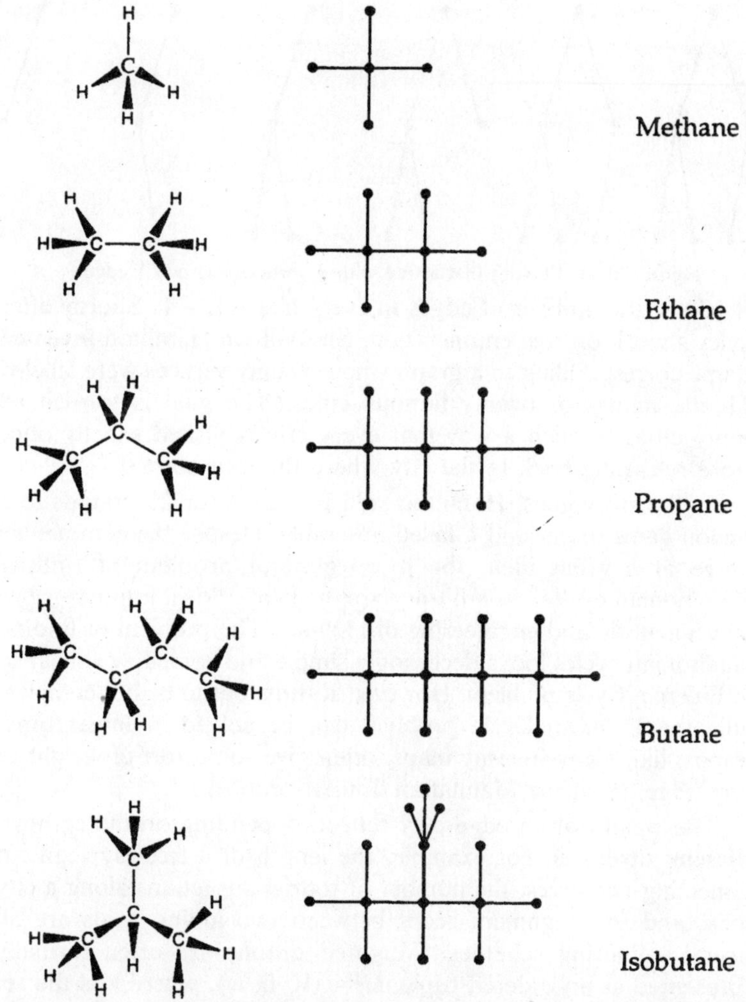

Figure 10.12. Hydrocarbons as chemists see them (left) and their graph representation (right). Two different molecules with the same number of the same types of atoms are called structural isomers.

Indeed, since every tree has a leaf, we can remove it and its attached edge, resulting in another tree. So far we have removed one edge and one vertex. In this smaller tree there exists a leaf that we, again remove. So far, we have removed two vertices and two edges. We keep this up until we are left with a graph with a single vertex and no edges. Since we have removed $n - 1$ vertices and n

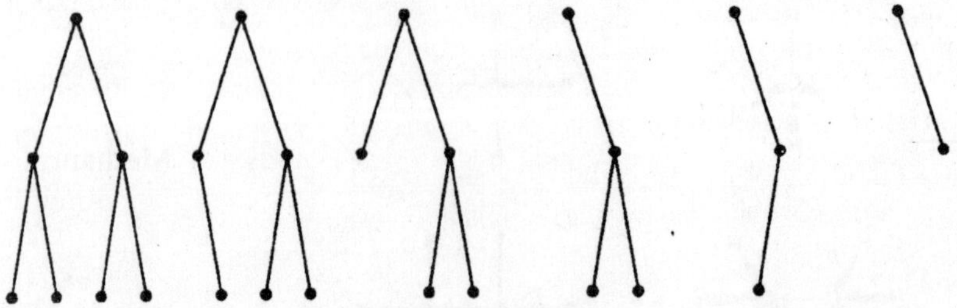

Figure 10.13. Proving that a tree with n vertices has n – 1 edges.

– 1 edges, the number of edges in every tree is $n - 1$. Shortly after Cayley's work on tree enumeration, Sir William Hamilton invented a game corresponding to a graph whose twenty vertices were labeled with the names of twenty famous cities. The goal is to visit all twenty cities in such a way that every city is visited exactly once before returning back to the city where the tour started.

As the story goes, Hamilton sold the game for 25 pounds to a London game dealer and it failed miserably. Despite the commercial failure of a great idea, the more general problem of finding "Hamiltonian cycles" in arbitrary graphs is of critical importance to many scientific and engineering disciplines. The problem of finding Hamiltonian cycles looks deceivingly simple and somehow similar to the Eulerian Cycle problem. However, it turns out to be NP-complete while the Eulerian Cycle problem can be solved in linear time. Graphs, like many freeway maps, often give some sort of weight to every edge, as in the Manhattan Tourist problem.

The weight of an edge may reflect, depending on the context, different attributes. For example, the length of a freeway segment connecting two cities, the number of tourist attractions along a city block, and the alignment score between two amino acids are all natural weighting schemes. Weighted graphs are often formally represented as an ordered triple, $G = (V, E, w)$, where V is the set of vertices in the graph, E is the set of edges, and w is a weight function defined for every edge e in E(i.e., $w(e)$ is a number reflecting the weights of edge e). Given a weighted graph, one may be interested in finding some shortest path between two vertices. Though this problem may sound difficult if you were given a complicated road map, it turns out that there exist fast algorithms to answer this question.

We emphasize that this problem is different—and somewhat more complicated—than the Longest Path in a Directed Acyclic

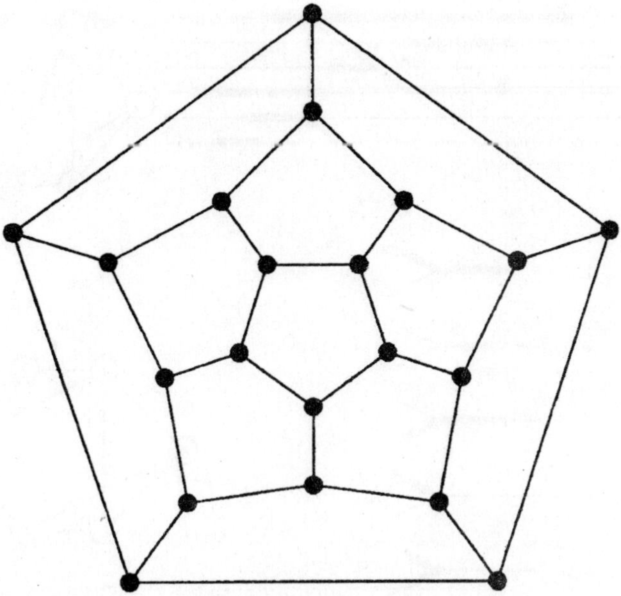

Figure 10.14. Can you travel from any one of the vertices in this graph, visit every other vertex exactly once, and end up at the original vertex.

Graph (DAG). Graphs are powerful tools in many applied studies and their role goes well beyond analysis of maps of freeways. For example, at first glance, the following "basketball" problem has nothing to do with graph theory: "Fifteen teams play a basketball tournament in which each team plays with each other team. Prove that the teams can always be numbered from 1 to 15 in such a way that team 1 defeated team 2, team 2 defeated team 3, .., team 14 defeated team 15."

A careful analysis reduces this problem to finding a Hamiltonian path in a directed graph on fifteen vertices.

10.5 GRAPH AND GENETICS

Conceived by Euler, Cayley, and Hamilton, graph theory flourished in the twentieth century to become a critical component of discrete mathematics. In the 1950s, Seymour Benzer applied graph theory to show that genes are linear. At that time, it was known that genes behaved as functional units of DNA, much like pearls on a necklace, but the chemical structures and organization of the genes was not clear. Were genes broken into still smaller components? If so, how were they organized?

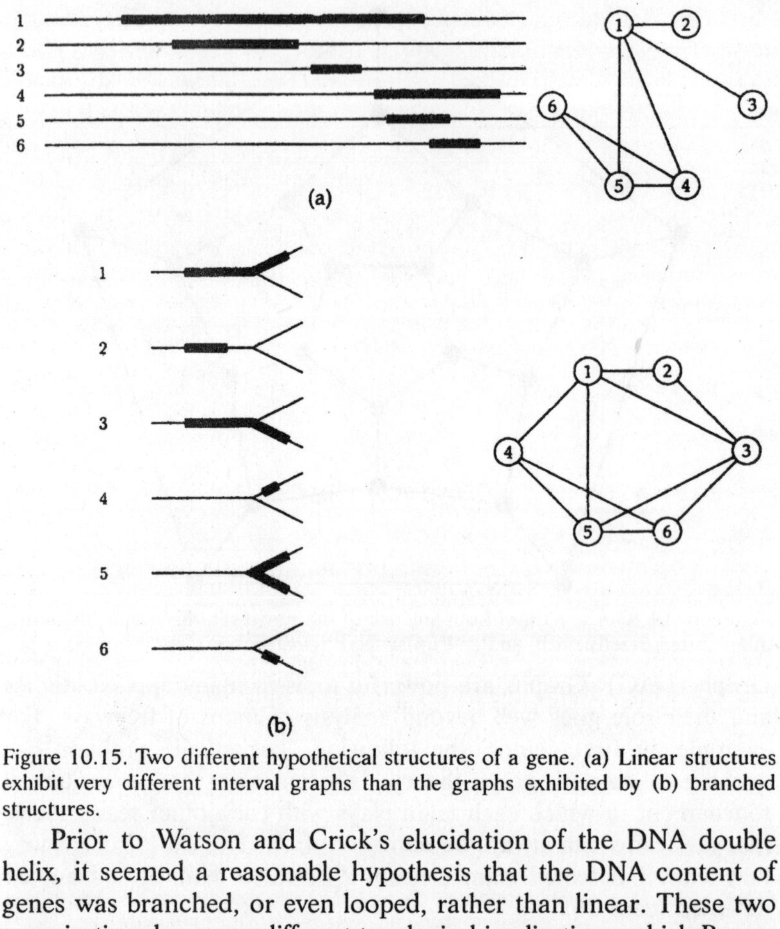

Figure 10.15. Two different hypothetical structures of a gene. (a) Linear structures exhibit very different interval graphs than the graphs exhibited by (b) branched structures.

Prior to Watson and Crick's elucidation of the DNA double helix, it seemed a reasonable hypothesis that the DNA content of genes was branched, or even looped, rather than linear. These two organizations have very different topological implications, which Benzer exploited in an ingenious experiment. Benzer studied a large number of bacteriophage T4 mutants, which happened to have a continuous interval deleted from an important gene. In their "normal" state, nonmutant T4 phages will kill a bacterium. The mutant T4 phages that were missing a segment of their genome could not kill the bacterium. Different mutants had different intervals deleted from the gene, and Benzer had to determine which interval had been deleted in each mutant.

Though Benzer did not know exactly where in the gene the mutant's deletion was, he had away to test whether two deletions (i.e., their intervals) overlapped, relying on how two phages with two different deletions behave inside a bacterial host cell. When two mutant phages with two different deletions infect a bacterium, two outcomes are possible depending on whether the deleted intervals overlap or not. If they do not overlap, then two phages combined have all the genetic material of one normal phage. However, if two deleted intervals overlap, some genetic material is absent in both phages.

The Benzer experiment was based on the observation that in the former case (all genetic material of a normal phage is present), two mutants are able to ill the bacterium; and in the later case (where some genetic material is removed in both phages) the bacterium survives. Benzer infected bacteria with each *pair* of mutant strains from his T4 phage and simply noted which pairs killed the bacterial host. Pairs which were lethal to the host were mutants whose deleted intervals did not overlap. Benzer constructed a graph where each strain was a vertex and two vertices were connected when a double infection of a bacterial host was nonlethal. He reasoned that if genes were linear, he would probably see one type of graph, but if the genes were branched he would see another.

Given a set of n intervals on a line segment, the *interval graph* is defined as a graph with n vertices that correspond to the intervals. There is an edge between vertices v and w if and only if the intervals v and w overlap. Interval graphs have several important properties that make them easy to recognize—for example, the graph in figure 10.16 (a) is an interval graph, whereas the "house" graph in figure 10.16 (b) is not. Benzer' problem was equivalent to deciding whether the graph obtained from his bacteriophage experiment represented an interval graph. Had the experiment resulted in a graph like the "house" graph, then the genes could not have been organized as linear structures. As it turned out, the graph was indeed an interval graph, indicating that genes were composed of linearly organized functional units.

10.6 DNA SEQUENCING

Imagine several copies of a magazine cut into millions of pieces. Each copy is cut in a different way, so a piece from one copy may overlap pieces from another. Assuming that some large number of pieces are just sort of lost, and the remaining pieces are splashed with ink, can you recover the original text? This essentially, is the problem of fragment assembly in DNA sequencing. Classic DNA sequencing technology allows one to read short 500- to 700-

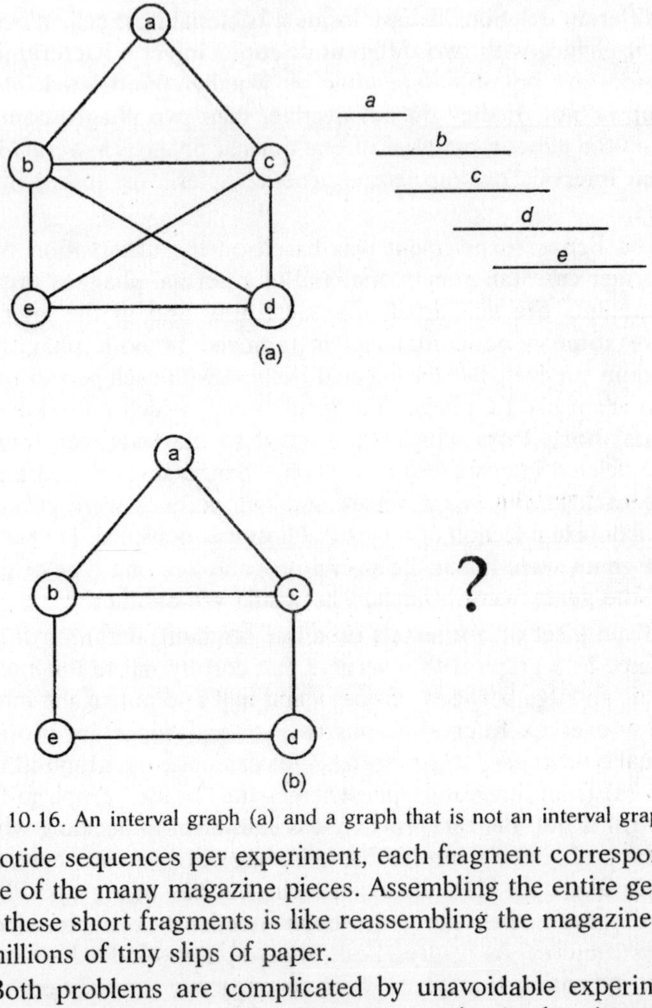

Figure 10.16. An interval graph (a) and a graph that is not an interval graph (b).

nucleotide sequences per experiment, each fragment corresponding to one of the many magazine pieces. Assembling the entire genome from these short fragments is like reassembling the magazine from the millions of tiny slips of paper.

Both problems are complicated by unavoidable experimental errors—ink splashes on the magazine, and mistakes in reading nucleotides. Furthermore, the data are frequently incomplete–some magazine pieces get lost, while some DNA fragments never make it into the sequencing machine. Nevertheless, efforts to determine the DNA sequence of organisms have been remarkably successful, even in the face of these difficulties. Two DNA sequencing methods were invented independently and simultaneously in Cambridge, England by Fred Sanger and in Cambridge, Massachusetts by Walter Gilbert. Sanger's method takes advantage of how cells make copies of DNA.

Cells copy a strand of DNA nucleotide by nucleotide in a reaction that adds one base at a time.

Sanger realized that he could make copies of DNA fragments of different lengths if he starved the reaction of one of the four bases: a cell can only copy its DNA while it has all of the bases in supply. For a sequence ACGTAAGCTA, starving at T would produce a mixture of the fragments ACG and ACGTAAGC. By running one starvation experiment for each of A, T, G, and C and then separating the resulting DNA fragments by length, one can read the DNA sequences. Each of four starvation experiments produces a *ladder* of fragments of varying lengths called the *Sanger ladder*. This approach culminated in the sequencing of a 5386-nucleotide virus in 1977 and a Nobel Prize shortly thereafter.

Since then the amount of DNA sequence data has been increasing exponentially, particularly after the launch of the Human Genome Project in 1989. By 2001, it had produced the roughly 3 billion-nucleotide sequence of the human genome. Within the past twenty years, DNA sequencing technology has been developed to the point where modern sequencing machines can sequence 500- to 700-nucleotide DNA fragments, called *sequencing reads*. These reads then have to be assembled into a continuous genome, which turns out to be a very hard problem.

Even though the DNA reading process has become quite automated, these machines are not microscope-like devices that simply scan 500 nucleotides as if they were a sentence in a book. The DNA sequencing machine measure the lengths of DNA fragments in the Sanger ladder, but even this task is difficult; we cannot measure a *single* DNA fragment, but must measure billions of identical fragments. *Shotgun sequencing* starts with a large sample of genomic DNA. The sample is *sonicated*, a process which randomly partitions each piece of DNA in the sample into *inserts*; the inserts that are smaller than 500 nucleotides are removed from further consideration. Before the inserts can be read, each one must be multiplied billions of times so that it is possible to read the ladders produced by Sanger's technique.

To amplify the inserts, a sample is cloned into a *vector*, and this vector used to infect a bacterial host. As the bacterium reproduces, it creates a colony that contains billions of copies of the vector and its associated insert. As a result, the cloning process results in the production of a large sample of one particular insert that can then be sequenced by the Sanger method. Usually, only the first 500 to 700 nucleotides of the insert can be interpreted from this experiment. DNA sequencing is therefore a two stage process including both experimental (reading 500-700 nucleotide sequences from different inserts) and computational (assembling these reads into a single long sequence) components.

10.7 SHORTEST SUPERSTRING PROBLEM

Since every string, or *read*, that we sequence came from the much longer genomic string, we are interested in a *superstring* of the reads—that is, we want a long string that "explains" all the reads we generated. However, there are many possible superstrings to choose from—for example, we could concatenate all the reads together to get a (not very helpful) superstring. We choose to be most interested in the shortest one, which turns out to be a reasonable first approximation to the unknown genomic DNA

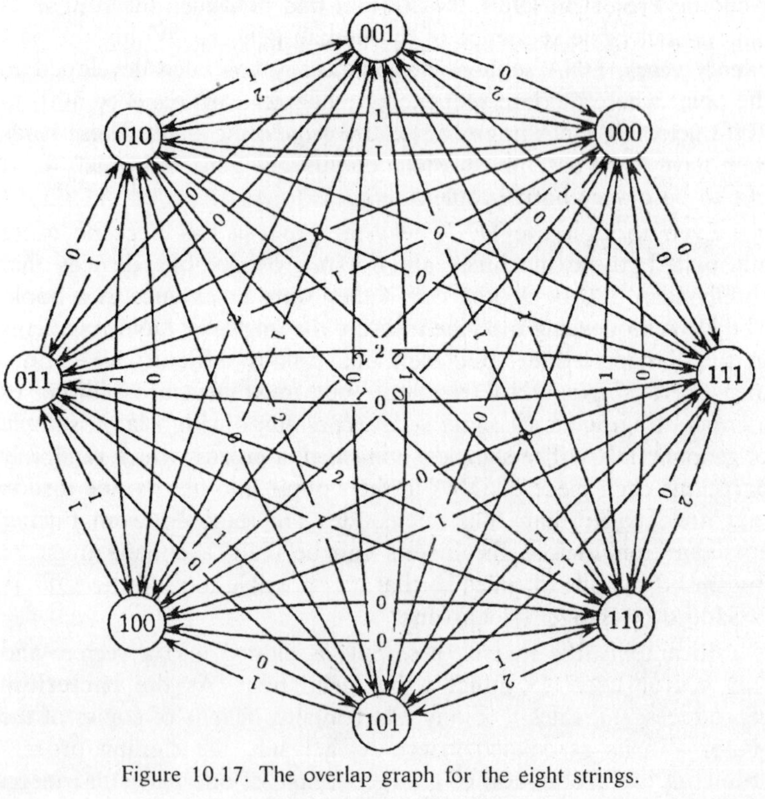

Figure 10.17. The overlap graph for the eight strings.

sequence. With this in mind, the simplest approximation of DNA sequencing corresponds to the following problems. Define *overlap*(s_i, s_j) to be the length of the longest prefix of s_j that matches a suffix of s_i.

The shortest superstring problem can be cast as a Traveling Salesman problem in a complete directed graph with n vertices corresponding to strings $s_1, ..., s_n$ and edges of length –*overlap*(s_i, s_j). This reduction, of course, does not lead to an efficient algorithm since the TSP is *NP*-complete. Moreover it is known that shortest superstring problem is itself *NP*-complete, so that a polynomial algorithm for this problem is unlikely. The early DNA sequencing algorithms used a simple greedy strategy: repeatedly merge a pair of strings with maximum overlap until only one string remains. It has been conjectured but not yet proved, that this greedy algorithms has performance guarantee 2.

10.8 DNA ARRAYS AS AN ALTERNATIVE SEQUENCING TECHNIQUE

When the Human Genome Project started, DNA sequencing was a routine but time-consuming and hard-to-automate procedure. In 1988 four groups of biologists independently and simultaneously suggested a different sequencing technique called *Sequencing by Hybridization*. SBH involves building a miniature *DNA array*, also known as *DNA chip*, that contains thousands of short DNA fragments called *probes*. Each of these short fragments reveals. Whether or not a known—but short—sequence occurs in the unknown DNA sequence; all these pieces of information together should reveal the identify of the target DNA sequence. Given a short probe (an 8- to 30-nucleotide single-stranded synthetic DNA fragments) and a single-stranded target DNA fragment, the target will *hybridize* with the probe if the probe is a substring of the target's Watson-Crick *complement*.

When the probe and the target are mixed together, they form a weak chemical bond and stick together. For example, a probe ACCGTGGA will hybridize to a target CCTGGCACCTA since its is complementary to the substring TGGCACCT of the target. In 1988 almost nobody believed that the idea of using DNA probes to sequence long genomes would work, because both the biochemical problem of synthesizing thousands of short DNA fragments and the combinatorial problem of sequence reconstruction appeared too complicated. Shortly after the first paper describing DNA arrays was published, the journal *Science* wrote that given the amount of work required to synthesize a DNA array, "[using DNA arrays for sequencing] would simply be substituting one horrendous task for another."

A major breakthrough in DNA array technology was made by Steve Fodor and colleagues in 1991. Their approach to array manufacturing relies on *light-directed polymer synthesis*, which has many similarities to computer chip manufacturing. Using this technique, building an array with all 4^l probes of length l requires

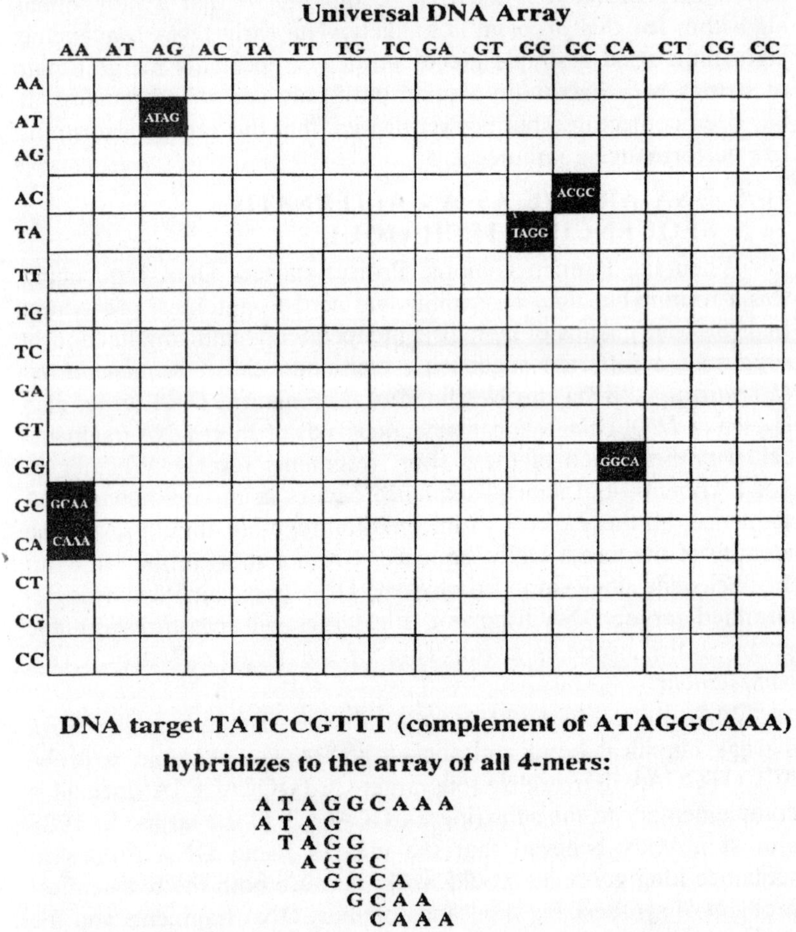

Figure 10.18. Hybridization of TATCCGTTT with the universal DNA array consisting of all 4^4 4-mers.

just $4 \cdot l$ separate reactions, rather than the presumed 4^l reactions. With this method, a California-based biotechnology company, Affymetrix, built the first 64-kb DNA array in 1994.

Today, building 1-MB or larger arrays is routine, and the use of DNA arrays has become one of the most widespread new biotechnologies. SBH relies on the hybridization of the target DNA fragment against a very large array of short probes. In this manner, probes can be used to test the unknown target DNA to determine its *l-mer composition*. The *universal* DNA array contains all 4^l probes of length l and is applied as follows:

1. Attach all possible probes of length l ($l = 8$ in the first SBH papers) to a flat surface, each probe at a distinct and known location. This set of probes is called the *DNA array*.
2. Apply a solution containing fluorescently labeled DNA fragment to the array.
3. The DNA fragment hybridizes with those probes that are complementary to substrings of length l of the fragment.
4. Using a spectroscopic detector, determine which probes hybridize to the DNA fragment to obtain the l-mer composition of the target DNA fragment.
5. Apply the combinatorial algorithm described below to reconstruct the sequence of the target DNA fragment from the l-mer composition.

10.9 SEQUENCING BY HYBRIDIZATION

Given an unknown DNA sequencing, an array provides information about all strings of length l that the sequence contains, but does not provide information about their positions in the sequence. For a string s of length n, the l-mer composition, or *spectrum*, of s, is the multiset of $n - l + 1$ l-mers in s and is written *Spectrum* (s, l). If $l = 3$ and s = TATGGTGC, then *Spectrum* (s, l) = {TAT, ATG, TGG, GGTS, GTG, TGC}. We can now formulate the problem of sequencing a target DNA fragment from its DNA array data. Although conventional DNA sequencing and SBH are very different experimental approaches, you can see that the corresponding computational problems are quite similar.

In fact, SBH is a particular case of the Shortest Superstring problem when the strings s_1, ..., s_n represent the set of all sub strings of s of fixed size. However, in contrast to the Shortest Superstring problem, there exists a simple linear-time algorithm for the SBH problem. Notice that it is not a contradiction that the shortest superstring problem is *NP*-complete, yet we claim to have a linear-time algorithm for the SBH problem, since the Shortest Superstring problem is more general than the SBH problem. Although DNA arrays were originally proposed as an alternative to conventional DNA sequencing, do novo sequencing with DNA arrays remains an unsolved problem in practice.

The primary obstacle to applying DNA arrays for sequencing is the inaccuracy in interpreting hybridization data to distinguish between perfect matches [i.e. l-mers present in *Spectrum*(s, l)] and highly stable mismatches (i.e. l-mers not present in *Spectrum*(s, l), but with sufficient chemical bonding potential to generate a strong hybridization signal). This is a particularly difficult problem for the short probes used in universal arrays. As a result, DNA arrays have become more popular in gene expression analysis and studies of genetic variations—both of which are done with longer probes—than in *de novo* sequencing.

In contrast to SBH where the target DNA sequence is unknown, these approaches assume that the DNA sequence is either known or "almost" known (i.e, known up to a small number of mutations). For example to detect genetic variations one can design twenty-to thirty-nucleotide probes to reliably detect mutations, bypassing the still unsolved problem of distinguishing perfect matches from highly stable mismatches in the case of short probes. To detect mutations in the (known) sequence S, an array should contain all 25-mers from S, as well a selected mutated versions of these 25-mers.

10.10 SBH AS A HAMILTONIAN PATH PROBLEM

Two l-mers p and q *overlap* if *overlap*$(p, q) = l - 1$, that is, the last $l - 1$ letters of p coincide with the first $l - 1$ letters of q. Given the measured spectrum *Spectrum*(s, l) of a DNA fragment s, construct a directed graph, H, by introducing a vertex for every l-mer in *Spectrum*(s, l), and connect every two vertices p and q by the directed edge (p, q) if p and q overlap. There is a one-to-one correspondence between paths that visit each vertex of H exactly once and DNA fragments with the spectrum *Spectrum*(s, l). The spectrum corresponds to the sequence reconstruction ATGCAGGTCC, which is the only path visiting all vertices of H:

ATG → TGC → GCA → CAG → AGG → GGT → GTC → TCC

The spectrum yields a more complicated graph with two Hamiltonian path, each path corresponding to two possible reconstruction: ATGCGTGGCA and ATGGCGTGCA. As the overlap graph becomes larger, this approach ceases to be practically useful since the Hamiltonian Path problem is *NP*-complete.

10.11 SBH AS AN EULERIAN PATH PROBLEM

As we have seen, reducing the SBH problem to a Hamiltonian Path problem does not lead to an efficient algorithm. Fortunately, reducing SBH to the *Eulerian Path* problem in a directed graph, which leads to the simple linear-time algorithm for sequence reconstruction mentioned earlier. The reduction of the SBH problem to an Eulerian Path problem is to construct a graph whose edges—rather than vertices—correspond to l-mers from *Spectrum*(s, l) and then to find a path in this graph visiting every edge exactly once. In this approach we build a graph G on the set of all $(l - 1)$-mers, rather than on the set of all l-mers as in the previous section. An $(l - 1)$ mer v is joined by a directed edge with an $(l - 1)$-mer w if the spectrum contains an l-mer for which the first $l - 1$ nucleotides coincide with v and the last $l - 1$ nucleotides coincide with w. Each l-mer from the spectrum corresponds to a directed edge in G rather than to a vertex as it does in H. Therefore, finding a DNA fragment containing all l-mers from the spectrum corresponds to finding a path visiting all *edges* of G, which is the problem of finding an Eulerian path.

Superficially finding an Eulerian path looks just as hard as finding a Hamiltonian path, but, as we show below, finding Eulerian paths turns out to be simple. We will first consider *Eulerian cycles,* that is, Eulerian paths in which the first and the last vertices are the same. A directed graph G is *Eulerian* if it contains an Eulerian cycle. A vertex v in a graph is *balanced* if the number of edges entering v equals the number of edges leaving v, that is, if *indegree*(v) = *outdegree*(v) For any given vertex v is an Eulerian graph, the number of times the Eulerian cycle enters v is exactly the same as the number of times it leaves v. Thus, is, if *indegree*(v) = *outdegree*(v) for every vertex v is an Eulerian graph, motivating the following theorem characterizing Eulerian graphs.

Theorem. A connected graph is Eulerian if the an only if each of it vertices is balanced.

Proof. First, it si easy to see that if a graph is Eulerian, then each vertex must be balanced, We show that if each vertex is a connected graph is balanced, then the graph is Eulerian.

To construct an Eulerian cycle, we start from an arbitrary vertex v and form any arbitrary path by traversing edges that have not

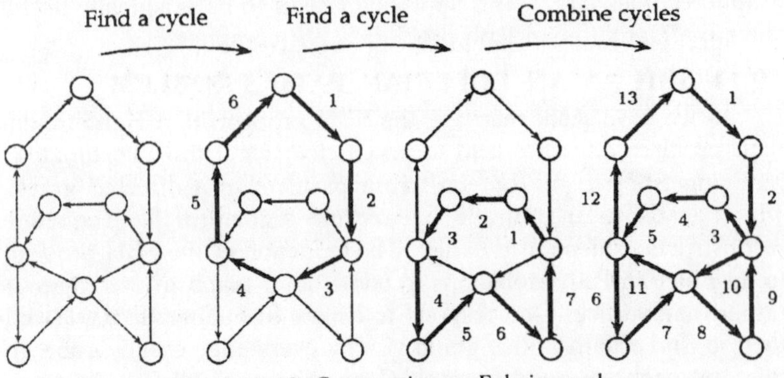

Figure 10.19. Constructing an Eulerian cycle.

already been used. We stop the path when we encounter a vertex with no way out, that is, a vertex whose outgoing edges have already been used in the path. In a balanced graph, the only vertex where this can happen is the starting vertex v since for any other vertex, the balance condition ensure that for every incoming edge there is an outgoing edge that has not yet been used. Therefore, the resulting path will end at the same vertex where it started, and with some luck will be Eulerian. However, if the path is not Eulerian, it must contain a vertex w that still has some number of untraversed edges.

The cycle we just constructed forms a balanced subgraph. Since the original graph was balanced, then the edges that were not traversed in the first cycle also form a balanced sub-graph. Since all vertices in the graph with untraversed edges are balanced there must exist some other path starting and ending at w, containing only untraversed edges. One can now combine the two paths into a single one as follows. Traverse the first path from v to w, then traverse the second path from w back to itself, and then traverse the remainder of the first path w back to v. Repeating this until there are no more vertices with unused edges will eventually yield an Eulerian cycle. This algorithm can be implemented in time linear in the number of edges in the graph.

Notice that we have described Eulerian graphs as containing an Eulerian *cycle*, rather than an Eulerian path, but we have said that the SBH problem reduces to that of finding an Eulerian path. A vertex v in a graph is called *semibalanced* if $|indegree(v) - outdegree(v)| = 1$. If a graph has an Eulerian path starting at vertex s and ending at vertex t, then all its vertices are balanced, with the possible exception of s and t, which may be semibalanced.

The Eulerian path problem can be reduced to the Eulerian cycle problem by adding an edge between two semibalanced vertices. This transformation balances all vertices in the graph and therefore guarantees the existence of an Eulerian cycle in the graph with the added edge. Removing the added edge form the Eulerian cycle transforms it into an Eulerian path. The following theorem characterizes all graphs that contain Eulerian paths.

Theorem. A connected graph has an Eulerian path if and only if it contains at most two semibalanced vertices and all other vertices are balanced.

10.12 FRAGMENT ASSEMBLY IN DNA SEQUENCING

As we mentioned previously, after the short 500- to 700-bp *DNA reads* are sequenced, biologists need to assemble them together to reconstruct the entire genomic DNA sequence. This is known as fragment assembly. The shortest superstring problem described above is an overly simplified abstraction that does not adequately capture the essence of the fragment assembly problem, since it assumes error-free read. The error rate in DNA reads produced by modern sequencing machines varies from 1% to 3%. A further complication in fragment assembly in the fact that one does not know a priori which of two DNA strands a read came from. DNA is double-stranded, and which of the two strands was sequenced by a read depends on how the insert was oriented in the vector. Since this is essentially arbitrary, one never knows whether a read came from a target strand DNA sequence or from its Watson-Crick complement. However, sequencing errors and assignments of reads to one of two strands are just minor annoyances compared to the major problem in fragment assembly: repeats in DNA.

The human genome contains many sequences that repeat themselves throughout the genome a surprisingly large number of times. For example, the roughly 300 nucleotide *Alu* sequence is repeated more than a million times throughout the genome, with only 5% to 15% sequence variation. Even more troublesome for fragment assembly algorithms is the fact that repeats occur at several scales. The human T-cell receptor locus contains five closely located repeats of the trypsinogen gene, which is 4 kb long and varies only by 3% to 5% between copies. These long repeats are particularly difficult to assemble, since there are no reads with unique portions flanking the repeat region. The human genome contains more than 1 million *Alu* repeats (≈ 300 bp) and 200,000 *LINE* repeats (≈ 100 bp), not to mention that an estimated 25% of genes in the human genome have duplicated copies.

A little arithmetic shows that these repeats and duplicated genes represent about half the human genome. If one models the genome as a 3 billion-letter sequence produced by a random number generator, then assembling it from 500-letter reads is actually relatively simple. However because of the large number of repeats, it is theoretically impossible to uniquely assemble real reads as long as some repeats are longer than the typical read length. Increasing the length of the reads (to make them longer than most repeats) would solve the problem, but the sequencing technology has not significantly improved the read length yet. A puzzle that looks deceivingly simple and has only sixteen triangular pieces. People usually assemble puzzle by connecting matching pieces. In this case, for every triangle in the puzzle, there is variety of potentially matching triangles (every frog in the puzzle is repeated several times). As a result, you cannot know which of the potentially matching triangles is the correct one to use at any step. If you proceed without some sort of guidance, you are likely to end up in the situation. Fourteen of the pieces have been placed completely consistently, but the two remaining pieces are impossible to place.

It is difficult to design a strategy that can avoid such dead ends for this particular puzzle, and it is even more difficult to design strategies for the linear puzzle presented by repeats in a genome. Since repeats present such a challenge in assembling long genomes, the original strategy for sequencing the human genome was first to clone it into BACs, each BAC carrying an approximately 150,000 bp long insert. After constructing a library of overlapping BACs that covers the entire human genome (which requires approximately 30,000 BACs), each one can be sequenced as if it were a separate minigenome. This BAC-by-BAC sequencing strategy significantly simplifies the computational assembly problem (by virtue of the fact that the number of repeats present within a BAC is 30,000 times smaller than the number of repeats in the entire genome) but makes the sequencing project substantially more cumbersome.

Although the Human Genome project demonstrated that this BAC-by-BAC strategy can be successful, recent large-scale sequencing projects (including the 2002 mouse genome assembly) mainly follow the whole-genome assembly paradigm advocated by James Weber and Gene Myers in 1997. Myers led the fragment assembly efforts at Celera Genomics, the company that announced the completion of a (draft) human genomic sequence in 2001. Weber and Myers suggested a virtual increase in the length of a read, by pairing reads that were separated by a fixed-size gap. This suggesting resulted in the so-called *mate-pair reads* sequencing technique. In this method, inserts of length approximately L (where L is much longer than the length of a read) are selected, an *both* ends of the insert are sequenced. This produces a pair of reads called mates at a known (approximate) distance, L, from each other. The insert length L is chosen in a such away that it is larger than the length of most repeats in the human genome. The advantage of mate-pair reads is that it is unlikely that both reads of the mate-pair will lie in a large-scale DNA repeat. Thus the read that lies in a unique portion of DNA determines which copy of a repeat its mate is in. Most fragment assembly algorithms consist of the following three steps:

- *Overlap*: Finding potentially overlapping reads.
- *Layout*: Finding the order of reads along DNA.
- *Consensus*: Deriving the DNA sequence form the layout.

The overlap problem is to find the best match between the suffix of one read and the prefix of another. In the absence of sequencing errors, we could simply find the longest suffix of one string that exactly matches the prefix of another string. However, sequencing errors force us to use a variation of the dynamic programming algorithm for sequence alignment. Since errors are small (1% to 3%), the common practice is to filter out pairs of fragments that do not share a significantly long common substring. Constructing the layout is the hardest step in fragment assembly. The difficulty is in deciding whether two fragments really overlap (i.e., their differences are caused by sequencing errors) or actually come from two different copies of repeat.

Repeats represent a major challenge for whole-genome shotgun sequencing and make the layout problem very difficult. The final consensus step of fragment assembly amounts to correcting errors in sequence reads. The simplest way to build the consensus is to report the most frequent character in the layout constructed in the layout step. This assumes that each position in the genome was represented by a sufficiently large number of reads to ensure that experimental errors are reduced to minor noise.

10.13 PROTEIN SEQUENCING AND IDENTIFICATION

Few people remember that before DNA sequencing had even been seriously suggested, scientists routinely sequenced proteins.

Frederick Sanger was awarded his first (of two) Nobel prize for determining the amino acid sequence of insulin, the protein needed by people suffering from diabetes. Sequencing the 52-amino acid bovine insulin in the late 1940s seemed more challenging than sequencing an entire genome seems today. The computational problem facing protein sequencing at that time was similar to that facing modern DNA sequencing; the main difference was in the length of the sequenced fragments.

In the late 1940s, biologists discovered how to apply the *Edman degradation reaction* to chop off one terminal amino acid at a time

```
GIVE
GIVEECCA
GIVEECCASV
GIVEECCASVC
GIVEECCASVCSL
GIVEECCASVCSLY
        SVC
           SLY
           SLYELEDYC
             YE
             YEL
             YELE
              ELEDY
              ELEDYCD
               LE
               LEDYCD
                EDYCD
                 DYCD
                  CD

                      FVDEHLCG
                      FVDEHLCGSHL
                          HLCGSHL
                              SHLVEA
                                 VEAL
                                 VEALY
                                   AL
                                   ALY
                                     YLVCG
                                      LVCGERGF
                                      LVCGERGFF
                                         GERG
                                            GF
                                            GFFYTPK
                                               YTPKA
                                                TPKA
```

Figure 10.20. The peptide fragments that Frederick Sanger obtained from insulin through a variety of methods.

from the end of a protein and read it. Unfortunately, this only works for a few terminal amino acids before the results become impossible to interpret. To get around this problem, Sanger digested insulin with *proteases* (enzymes that cleave proteins) into *peptides* (short protein fragments) and sequenced each of the resulting fragments independently. He then used these overlapping fragments to reconstruct the entire sequence, exactly like the DNA sequencing "break—read the fragments—assemble" method today.

The Edman degradation reaction became the predominant protein sequencing method for the next twenty years, and by the late 1960s, protein sequencing machines were on the market. Despite these advances, protein sequencing ceased to be of central interest in the field as DNA sequencing technology underwent rapid improvement in the late 1970s. In DNA sequencing, obtaining reads is relatively easy; it is the assembly that is difficult. In protein sequencing, obtaining reads is the primary problem, while assembly is easy. Having DNA sequence data for a cell is critical to understanding the molecular processes that the cell goes through. However, it is not the only important component: one also needs to know what proteins the cell produces and what they do. On the one hand, we do not yet know the full set of proteins that cells produce, so we need way to discover the sequence of previously unknown proteins. On the other hand, it is important to identify which specific proteins interact in a biological system (e.g., these proteins involved in DNA replication).

Lastly, different cells in an organism have different repertoires of expressed proteins. Brain cells need different proteins to function than liver cells do, and an important problem is to identify proteins that are present or absent in each biological tissue under different conditions. There are two types of computational problems motivated by protein sequencing. *De novo protein sequencing* is the elucidation of a protein's sequence in the case when a biological sample contains a protein that is either not present in a database or differs from a canonical version present in a database (e.g., mutated proteins or proteins with biochemical modifications).

The other problem is the identification of a protein that is present in a database; this is usually referred to as *protein identification*. The main difference between protein sequencing algorithms and protein identification algorithms is the difficulty of the underlying computational problems. Perhaps the easiest way to illustrate the distinction between protein identification and sequencing is with a gedanken experiment. Suppose a biologist wants to determine which proteins from the DNA polymerase complex in rats. Having the complete rate genome sequence and knowing the location of all the rat genes does not yet allow a biologist to determine what chemical reactions occur during the DNA replication process. However, isolating a rat's DNA polymerase complex, breaking it apart, and sequencing the proteins that form parts of the complex will yield a fairly direct answer to the researcher's question.

Of course, if we presume that the biologist has the complete rat genome sequence and all of its gene *products*, he may not actually have to sequence every amino acid in every protein in the DNA polymerase complex—just enough to figure out which proteins are present. This is protein identification. On the other hand, if the researcher decides to study an organism for which complete genome data are not available (perhaps an obscure species of ant), then the researcher will need to perform de novo protein sequencing. For many problems, protein sequencing and identification remains the only ways to probe a biological process. For example gene splicing is a complex process performed by the large molecular complex called the *spliceosome*, which consists of over 100 different proteins complexed with some functional RNA. Biologists want to determine the "parts list" of the spliceosome, that is, the identity of proteins that form the complex. DNA sequencing is not capable of solving this problem directly: even if all the proteins in the genome were known, it is not clear which of them are parts proteins in the genome were known, it is not clear which of them are parts of the spliceosome.

Protein sequencing and identification, on the other hand, are very helpful in discovering this parts list. Recently, Matthias Mann and colleagues purified the spliceosome complex and used protein sequencing and protein identification techniques to find a detailed parts list for it. Another application of these technologies is the study of proteins involved in *programmed cell death*, or *apoptosis*. In the development of many organisms cells must die at specific times. A cell dies if it fails to acquire certain survival factors, and the death process can be initiated by the expression of certain genes. In a developing nematode, for example, the death of individual cells in the nervous system may be prevented by mutations in several genes that are the subject of active investigation. DNA sequence data alone are not sufficient to find the genes involved in programmed cell death, and until recently, nobody knew the identity of these proteins.

Protein analysis by *mass spectrometry* allowed the sequencing of proteins involved in programmed cell death, and the discovery of some proteins involved in the death-including signaling complex. The exceptional sensitivity of mass spectrometry has opened up new experimental and computational possibilities of protein studies. A protein can be digested into peptides by proteases like trypsin. In a matter of seconds, a *tandem mass spectrometer* breaks a peptide into even smaller fragments and measures the mass of each. The *mass spectrum* of a peptide is a collection of masses of these fragments.

The protein sequencing problem is to derive the sequence of a peptide given its mass spectrum. For an ideal fragmentation process where every fragment of a peptide is generated, an in an ideal mass spectrometer, the peptide sequencing problem is simple. However, the fragmentation process is not ideal, and mass spectrometers measures mass with some imprecision. These details make peptide sequencing difficult. A mass spectrometer works like a charged sieve. A large molecule (peptide) gets broken into smaller fragments that have an electrical charge. These fragments are then spun around and accelerated in a magnetic field until they hit a detector. Because large fragments are harder to spin than small ones, one can distinguish between fragments with different masses based on the amount of energy required to fling the different fragments around. It happens that most molecules can be broken in several places, generating several different *ion types*. The problem is to reconstruct the amino acid sequence of the peptide from the masses of these broken pieces.

10.14 PEPTIDE SEQUENCING PROBLEM

Let $A = \{a_1, a_2, \ldots a_{20}\}$ be the set of amino acids, each with molecular masses $m(a_i)$. A *peptide* $P = p_1 \ldots p_n$ is a sequence of amino acids, with *parent* mass $m(P) = \Sigma^n_{i=1} m(p_i)$. We will denote the partial *N-terminal* peptide $p_1, \ldots p_i$ of mass $m_i = \Sigma^i_{j=1} m(p_j)$ as P_i and partial *C-terminal* peptide $p_{i+1}, \ldots, p_n$ of mass $m(P) - m_i$ as P_i^-, for $1 \leq i \leq n$. Mass spectra obtained by tandem mass spectrometry (MS/MS) consist predominantly or partial N-terminal peptides and C-terminal peptides.

A mass spectrometer typically breaks a peptide $p_1 p_2 \ldots p_n$ at different peptide bonds and detects the masses of the resulting partial

N-terminal and C-terminal peptides. For example, the peptide GPFNA may be broken into the N-terminal peptides G, GP, GPF, GPFN, and C-terminal peptides PFNA, FNA, NA, A. Moreover, while breaking GPFNA into GP and FNA, it may lose some small parts of GP and FNA, resulting in fragments of a lower mass. For example, the peptide GP might lose a water (H_2O), and the peptide FNA might lose an ammonia (NH_3). The resulting masses detected by the spectrometer will be equal to the mass of GP minus the mass of water (water happens to weight 1 + 1 + 16 = 18 daltons), and the mass of FNA minus the mass of ammonia (1 + 1 + 1 + 14 =17 daltons). Peptides missing water and ammonia are two different *ion types* that can occur in fragmenting a peptide in a mass spectrometer.

Peptides fragmentation in a tandem mass spectrometer can be characterized by a set of numbers $\Delta = \{\delta_1, ..., \delta_k\}$ representing the different types of ions that correspond to the removal of a certain chemical group from a peptide fragment. We will call Δ the set of ion types. A δ-*ion* of N-terminal partial peptide P_i is a modification of P_i that has mass $m_i = \delta$, corresponding to the loss of a (typically small) chemical group of mass δ when P was fragmented into P_i. The δ-*ion* of C-terminal peptides is defined similarly. The most frequent N-terminal ions are called b-ions (ion b_i corresponds to P_i with $\delta = -1$) and the most frequent C-terminal ions are called y-ions (ion y_i corresponds to P_i^- with $\delta = 19$). Examples of other frequent N-terminal ions are represented by b-H_2O (a b-fragment that loses a water) or y-NH_3 and some others like b-H_2O-NH_3.

For tandem mass spectrometry, the *theoretical* spectrum $T(P)$ of peptides P can be calculated by subtracting all possible ion types $\delta_1, ..., \delta_k$ from the masses of all partial peptides of P, such that every partial peptide generates k masses in the theoretical spectrum. An *experimental* spectrum $S = \{s_1, ..., s_q\}$ is a set of numbers obtained in a mass spectrometry experiment that includes masses of some fragment ions as well as chemical noise. Note that the distinction between the theoretical spectrum $T(P)$ and the experimental spectrum S is that you mathematically generate $T(P)$ given the peptide sequence P, but you experimentally generate S without knowing what the peptide sequence is that generated it.

The *match* between the experimentally measured spectrum S and peptide P is the number of masses in S that are equal to masses in $T(P)$. This is often referred to as the *shared peaks count*.

In reality, peptide sequencing algorithms use more sophisticated objective functions than a simple shared peaks count, incorporating different weighting functions for the matching masses. We formulate the Peptide Sequencing problem as follows. In reality, mass spectrometers measure both mass and *intensity*, which reflects the number of fragment ions of a given mass are detected in the mass spectrometer. As a result, mass spectrometrists often represent spectra in two dimensions and refer to the masses in the spectrum as "peaks."

10.15 SPECTRUM GRAPHS

There are two main approaches to solving the Peptide Sequencing problem that researchers have tried: either through exhaustive search among all amino acid sequences of a certain length, or by analyzing the *spectrum graph* which we define below. The former approach involves the generation of all 20^l amino acid sequence of length l and their corresponding theoretical spectra, with the goal of finding a sequence with the best match between the experimental spectrum and the sequence's theoretical spectrum. Since the number of sequences grows exponentially with the length of the peptide, different branch-and-bound techniques have been designed to limit the combinatorial explosion in these methods.

Prefix pruning restricts the computational space to sequences whose prefixes match the experimental spectrum well. The difficulty with the prefix pruning is that it frequently discards the correct sequence if its prefixes are poorly represented in the spectrum. The spectrum graph approach, on the other hand, does not involve generating all amino acid sequences, and leads to a fast algorithm for peptide sequencing. In this approach, we construct a graph from the experimental spectrum. Assume for simplicity that an experimental spectrum $S = \{s_1, ..., s_q\}$ consists of N-terminal ions and we will ignore the C-terminal ions for a while. Every mass $s \in S$ may have been created form a partial peptide by one of the k different ion types. Since we do not know which ion type from $\Delta = (\delta_1, ..., \delta_k)$ created the mass s in the experimental spectrum, we generate k different "guesses" for each of masses in the experimental spectrum.

Every guess corresponds to the hypothesis that $s = x - \delta_j$, where x is the mass of some partial peptide and $1 \leq j \leq k$. Therefore, for every mass s in the experimental spectrum, there are k guesses for the mass x of some partial peptide: $s + \delta_1$, $s + \delta_2$, ..., $s + \delta_k$. As a result, each mass in the experimental spectrum is transformed into a set of k vertices in the spectrum graph, one for each possible ion type. The vertex for δ_i for the mass s is labeled with mass $s + \delta_i$. We connect any two vertices u and v in the graph by the directed edge (u, v) if the mass of v is larger than that of u by the mass of a single amino acid. If we add a vertex at 0 and a vertex at the parent mass m (connecting them to other vertices as before), then the Peptide Sequencing problem can be cast as finding a path from 0 to m in the resulting DAG. In summary, the vertex set of the resulting spectrum graph is a set of numbers $s_i + \delta_j$ representing potential masses of N-terminal peptides adjusted by the ion type δ_j. Every mass s_i of spectrum S generates k distinct vertices $V_i(s) = \{s_i + \delta_1, ..., s_i + \delta_k\}$, though the sets V_i and V_j may overlap if s_i and s_j are close. The set of vertices in a spectrum graph is therefore $\{s_{initial}\} \cup V_1 \cup ... \cup V_q \cup \{s_{final}\}$, where $s_{initial} = 0$ and $s_{final} = \text{m}$.

The spectrum graph may have a most $q_k + 2$ vertices. We label the edges of the spectrum graph by the amino acid whose mass is equal to difference between vertex masses. If we look at vertices as putative N-terminal peptides, the edge form u to v implies that the N-terminal sequence corresponding to v may be obtained by extending the sequence at u by the amino acid that labels (u, v). A spectrum S of a peptide $P = p_1...p_n$ is called *complete* if S contains a least one ion type corresponding to every N-terminal partial peptide P_i for every $1 \leq i \leq n$. The use of a spectrum graph is based on the observation that for a complete spectrum there exists a path of length $n + 1$ from $s_{initial}$ to s_{final} in the spectrum graph that is labeled by P.

This observation casts the Peptide Sequencing problem as one of finding the "correct" path in the set of all paths between two vertices in a directed acyclic graph. If the spectrum is complete, then the correct path that we looking for is often the path with the maximum number of edges, the familiar Longest Path in a DAG problem. Unfortunately, experimental spectra are frequently incomplete. Moreover, even if the experimental spectrum is complete, there are often many paths in the spectrum graph to choose from that have same (or even larger) length, preventing one from unambiguously reconstructing the peptide.

The problem with choosing a path with a maximum number of edges is that it dose not adequately reflect the "importance" of different vertices. For example, a vertex in the spectrum graph obtained by a shift of +1 as $s_i + 1$ (corresponding to the most frequent b-ions) should be scored higher than a vertex obtained by a shift of the rate b_2-H_2O-NH_3 ion ($s_i + 1 - 18 - 17 = s_i - 34$). Further, whenever there are two peaks s_i and s_i' such that $s_i + \delta_j = s_i' + \delta_{j'}$, the vertex corresponding to that mass should also get a higher score than a vertex obtained by a single shift. In the probabilistic approach to peptide sequencing, each ion type δ_i has some probability of occurring which we write as $p(\delta_i)$. Under the simplest assumption, the probability that δ_i occurs for some partial peptide is independent of whether δ_j also occurs for the same partial peptide.

Under this assumption, any given partial peptide may contribute as many as k masses in the spectrum [this happens with probability $\Pi^k_{i=1} p(\delta_i)$] and as few as 0 [this happens with probability $\Pi^k_{i=1} (1 - p(\delta_i))$]. The probabilistic model below scores the vertices of the spectrum graph based on these simple assumptions. Suppose that an N-terminal partial peptide P_i with mass m_i produces ions $\delta_i, \ldots, \delta_l$ ("present" ions of mass $m_i - \delta_i, \ldots, m_i - \delta_l$) but fails to produce ions $\delta_l + 1, \ldots, \delta_k$ ("missing" ions) in the experimental spectrum. All l present ions will result in a vertex in the spectrum graph at mass m_i, corresponding to P_i. How should we score this vertex? A naive approach would be to reward P_i for every ions type that explains it, suggesting a score of $\Pi^l_{i=1} p(\delta_i)$. However, this approach has the disadvantage of not considering the missing ions, so we combine those in by defining the score of the partial peptide to be

$$\left(\prod_{i=1}^{l} p(\delta_i)\right)\left(\prod_{i=l+1}^{k} (1 - p(\delta_i))\right)$$

However, there is some inherent probability of chemical noise, that is, it can produce *any* mass (that has nothing to do with a peptide of interest) with certain probability p_R. Therefore, we adjust the probabilistic score as

$$\left(\prod_{i=1}^{i} \frac{p(\delta_i)}{p_R}\right)\left(\coprod_{i=l+1}^{k} \frac{1 - p(\delta_i)}{1 - p_R}\right)$$

10.16 PROTEIN IDENTIFICATION VIA DATABASE SEARCH

De novo protein sequencing algorithms are invaluable for identification of (both known and unknown) proteins, but they are most useful when working with complete or nearly complete high-quality spectra. Many spectra are far from complete, and de novo peptide sequencing algorithms often produce ambiguous solutions for such spectra. If we had access to a database of all proteins from a genome, then we would not longer need to consider all 20^l peptide sequences to interpret an MS/MS spectrum, but could instead limit our search to peptides present in this database.

Currently, most proteins are identified by database search—effectively looking the answer up in "the back of a book." Indeed, an experimental spectrum can be compared with the theoretical spectrum for each peptide in such a database, and the entry in the database that best matches the observed spectrum usually provides the sequence of the experimental peptides. This forms the basis of the popular SEQUEST algorithm developed by John Yates and colleagues. Though the logic that SEQUEST used to determine whether a database entry matches an experimental spectrum is somewhat involved, the basic approach of the algorithms is just a linear search through the database.

One drawback to MS/MS database search algorithms like SEQUEST is that peptides in a cell are often slightly different from the "canonical" peptides present in databases. The synthesis of proteins on ribosomes is not the final step in a protein's life: many proteins are subject to further modifications that regulate protein activities and these modifications may be either permanent or reversible. For example, the enzymatic activity of some proteins is regulated by the addition or removal of a phosphate group at a specific residue. *Phosphorylation* is a reversible process: *protein kinases* add phosphate groups while *phosphatases* remove them. Proteins form complex systems necessary for cellular signaling and metabolic regulation, and are therefore often subject to a large number of biochemical modifications (e.g., phosphorylation and glycosylation).

In fact, almost all protein sequences are modified after they have been constructed from their mRNA template, and as many as 200 distinct types of modifications of amino acid residues are known. Since we are unable to predict these *post-translational modifications* from DNA sequences, finding naturally occurring modifications remains an important open problem.

Computationally, a chemical modification of the protein $p_1p_2 \cdots p_i \cdots p_n$ at the position i results in increased mass of the N-terminal peptides $P_i, P_{i+1}, \ldots, P_n$ and increased mass of the C-terminal peptides $P_1^-, P_2^-, \ldots, P_{i-1}^-$. The computational analysis of modified peptides was also pioneered by John Yates, who suggested an exhaustive search approach that (implicitly) generates a virtual database of all modified peptides from a small set of potential modifications, and matches the experimental spectrum against this virtual database. It leads to a large combinatorial problem, even for a small set of modification types. The major difficulty with the Modified Protein Identification problem is that very similar peptides P_1 and P_2 may have very different spectra S_1 and S_2.

Our goal is to define a notion of spectral similarity that correlates well with sequence similarity. In other words, if P_1 and P_2 are a few modifications apart, the spectral similarity between S_1 and S_2 should be high. The shared peaks count is, of course, an intuitive measure of spectral similarity. However, this measure diminishes very quickly as the number of mutations increases, thus leading to limitations in detecting similarities by database search. Moreover, there are many correlations between the spectra of related peptides, and only a small proportion of these correlations is captured by the shared peaks count. The spectral convolution algorithm, below, reveals potential peptide modifications without an exhaustive search and therefore does not require generating a virtual database of modified peptides.

11

Computer in Molecular Studies

11.1 MOLECULAR EVOLUTION IS A FUNDAMENTAL PART OF BIOINFORMATICS

Genomes are dynamic molecular entities that evolve over time due to the cumulative effects of mutation, recombination, and selection. Before we address bioinformatics techniques for analysis of evolutionary relationships between biological sequences, and between protein structures, in later chapters, we will first survey these mechanisms that form the basis of genome evolution.

11.1.1 A Brief History of the Gene

The systematic study of the laws of heredity began with the work of Gregor Mendel (1822-1884). In 1865, Mendel who was an Augustinian monk living in Brno, then part of the Austro-Hungarian Empire, published a paper describing the results of plant-breeding experiments that he had begun in the gardens of his monastery almost a decade earlier. Mendel's work received little attention during his lifetime, but when his paper was rediscovered by biologists in 1900, a scientific revolution ensued. The key concept of the Mendelian revolution is that heredity is mediated by discrete units that can combine and dissociate in mathematically predictable ways.

Mendel had studied physics and plant physiology at the University of Vienna, where he was also introduced to the new science of statistics. His experimental methods show the mental habits of a physicist, and in looking for mathematical patterns of inheritance, he became the first mathematical biologist. He set out to arrange in a statistically accurate way the results of deliberate crosses between varieties of sweet pea plants with characteristics that could be easily identified. His elegant experiments led to a clear distinction between *genotype* (the hereditary make-up of an organism) and *phenotype* (the organism's physical and behavioural characteristics), and his results on the pattern of inheritance have become known as *Mendel's law.*

Mendel was the first to refer to hypothetical 'factors' that act as discrete units of heredity and are responsible for particular phenotypic traits. The rediscovery of his work at the beginning of the 20th century prompted a search for the cellular and molecular basis of heredity. The existence of 'germ plasm' was postulated, a material substance in eggs and sperm that in some way carried heritable traits from parent to offspring. While the molecular basis of hereditary factors—protein versus nucleic acid—remained in dispute until the mid-20th century, their cellular basis in chromosomes was soon discovered. In 1909, W.Johannsen coined the word *gene* to denote hypothetical particles that are carried on chromosomes and mediate inheritance.

In sexually reproducing *diploid* organisms, such as Mendel's pea plants and the fruit fly *Drosophila melanogaster* used in the breeding experiments of early Mendelians, the pattern of inheritance of some phenotypic traits could be explained by postulating a pair of genes underlying each trait—a pair of *alleles* occupying a *locus* on a chromosome. It was recognized early on that a single trait might be caused by several genes *(polygenic traits)* and that a single gene may have several effects *(pleiotropy).*

It is important to realize that 'gene' was an abstract concept to Mendel, and the Mendelian biologists of the first half of the 20th century. The founders of genetics, not having any knowledge of the biochemical basis of heredity, had to infer the characteristics of genes by observing the phenotypic outcomes of their breeding experiments. They developed a theoretical framework based on sound mathematics, now called *classical genetics,* that worked extremely well and is still useful today for the interpretation of the data obtained by the new molecular genetics.

Molecular genetics seeks to elucidate the chemical nature of the hereditary material and its cellular environment. At the beginning of the 1950s, it finally became clear that DNA was the critical ingredient of the genes. Rosalind Franklin (1920-1958), a crystallographer working at King's College in London, conducted a careful analysis of DNA using X-ray diffraction which indicated that the macromolecule possessed a double-stranded helical geometry. In 1953, James Watson and Francis Crick produced a successful model of the molecular structure of DNA based on her data by employing the molecular-model building method pioneered by Linus Pauling (1901-1994).

It was clear as soon as the structure of DNA was elucidated that its central role depends on the fact that it can be both replicated and read. This has given rise to the concept of *genetic information* being encoded in DNA. This idea is often expressed by metaphors that conceive of a gene as a 'word' and a genome, the total genetic material of a species, as a linguistic 'text' written in DNA code.

Over the ensuing five decades, rapidly accumulating knowledge on the fine structure of DNA and its functional organization has led to many critical adjustments and refinements in our understanding of the complex roles of genetic material in the cell. These discoveries all highlight the crucial contributions of the cellular environment in regulating the effects of DNA sequences on an organism's phenotype. The causal chain between DNA and phenotype is indirect, different cellular environments link identical DNA sequences to quite different phenotypic outcomes. One of the outcomes of the Human Genome Project is a significantly expanded view of the role of the genome within the integrated functioning of cellular systems.

The concept of the gene has undergone a profound transformation in recent years. For molecular biology, the traditional definition of gene action, originating from George Beadle's *one gene-one enzyme hypothesis* (1941) led to the concept of the gene as a stretch of DNA that codes for a single polypeptide chain. But even ignoring the fact that open reading frames might overlap, the relationship between DNA sequences and protein chains is many-to-many, not one-to-one. Now, the essential contributions of alternative splicing, RNA editing and post-translational modifications to the synthesis of the actual gene product have become recognized, and with them the limitations of the classic gene concept.

The gene as a unit of function can no longer be taken to be identical with the gene as a unit of intergenerational transmission of molecular information. New concepts are emerging that place emphasis on a *functional perspective* and define genes as 'segments of DNA that function as functional units', 'loci of cotranscribed exons' or, along similar lines, 'distinct transcription units or parts of transcription units that can be translated to generate one or a set of related amino acid sequences'. Another less widely adopted, yet, from a functional perspective, valid definition has recently been proposed by Eva Neumann-Held which states that a gene 'is a process that regularly results, at some stage in development, in the production of a particular protein.' This process centrally involves a linear sequence of DNA, some parts of which correspond to the protein via the genetic code. However, the concept of gene-as-process could also include such entities as coding regions for transcription factors that bind to its regulatory sequences, coding regions for RNA editing and splicing factors, the regulatory dynamics of the cell as a whole and the signals determining the specific nature of the final transcript, and beyond that, the final protein product. In conclusion, diverse interpretations of the concept of the gene exist and the meaning in which the term is applied can only be made clear by careful definition. At the very least, inclusion or exclusion of introns, regulatory regions and promoters need to be made explicit when we speak of a gene from the perspective of molecular biology.

11.1.2 What is Information?

Biological, or genetic, information is a fundamental concept of bioinformatics. Yet what exactly is information? In physics, it is understood as a measure of order that can be applied to any structure or system. The term 'information' is derived from the Latin *informare,* which means to 'form', to 'shape', to 'organize'. The word 'order' has its roots in textile weaving; it stems from the Latin *ordiri,* to 'lay the warp'. *Information theory,* pioneered by Claude Shannon, is concerned with information as a universal measure that can be applied equally to the order contained in a hand of playing cards, a musical score, a DNA or protein sequence, or a galaxy.

Information quantifies the instructions needed to produce a certain organization. Several ways to achieve this can be envisaged, but a particularly parsimonious one is in terms of binary choices. Following this approach, we compute information inherent in any given arrangement of matter from the number of 'yes' and 'no' choices that must be made to arrive at a particular arrangement among all equally possible ones. In his book *The Touchstone of Life* (1999), Werner Loewenstein illustrates this with the following thought experiment. Suppose you are playing bridge and are dealt a hand of 13 cards. There are about 635×10^9 different hands of 13 cards that can occur in this case, so the probability that you would be dealt this hand of cards is about 1 in 635×10^9—a large number of choices would need to be made to produce this exact hand. In other words, a particular hand of 13 cards contains a large amount of information.

Order refers to the structural arrangement of a system, something that is easy to understand in the case of a warp for weaving cloth but is much harder to grasp in the case of macromolecular structures. We can often immediately see whether an everyday structure is orderly or disorderly - but this intuitive notion does not go beyond simple architectural or periodic features. The intuition breaks down when we deal with macromolecules, like DNA, RNA and proteins. The probability for spontaneous assembly of such molecules is extremely low, and their structural specifications require enormous amounts of information since the number of ways they can be assembled as linear array of their constituent building blocks, nucleotides and amino acids, is astronomical. Like being dealt a particular hand during a bridge game, the synthesis of a particular biological sequence is very unlikely—its information content is therefore very high.

The large molecules in living organisms offer the most striking example of information density in the universe. In human DNA, roughly 3×10^9 nucleotides are strung together on the scaffold of the phosphate backbone in an aperiodic, yet perfectly determined, sequence. Disregarding spontaneous somatic mutations, all the DNA molecules of an individual display the same sequence. We are not yet able to precisely calculate the information inherent in the human genome, but we can get an idea of its huge information storing capacity with a simple calculation. The positions along the linear DNA sequence, that can be occupied by one of the four types of DNA bases, represent the elements of stored information. So, with 3×10^9 positions and four possible choices for each position, there are $4^{3,000,000,000}$ possible states. The number of possibilities is greater than the estimated number of particles in the universe.

Charles Darwin pointed out in *Origin* of *Species* how natural selection could gradually accumulate information about biological structures through the processes of random genotypic variation, natural selection and differential reproduction. What Darwin did not know was exactly how this information is stored, passed on to offspring and modified during evolution - this is the subject of the study of *molecular evolution*.

11.1.3 Molecular Evolution

This section provides an overview of the processes giving rise to the dynamics of evolutionary change at the molecular level. It primarily focuses on biological mechanisms relevant to the evolution of genomic sequences encoding polypeptide chains.

11.1.3.1 The algorithmic nature of molecular evolution

In his theory of evolution, Darwin identified three major features of the process that occur in an endlessly repeating cycle: generation of heritable variation by random mutation at the level of the genotype, natural selection acting on the phenotype, and differential reproductive success. Darwin discovered the power of cumulative algorithmic selection, although he lacked the terminology to describe it as such.

Algorithm is a computer science term and denotes a certain kind of formal process consisting of simple steps that are executed repetitively in a defined sequential order and will reliably produce a *definite kind of result* whenever the algorithm is run. Knitting is an algorithmic process—following the instructions 'knit one-purl one' will result in a particular ribbing pattern; other examples include braiding hair, building a car on an assembly line, or protein synthesis. Evolution by means of natural selection can also be thought of as algorithmic (mutate-select-replicate).

One important feature of an algorithm is its substrate neutrality: The power of the procedure is due to its logical structure, not the materials used in its instantiation, as long as the materials permit the prescribed steps to be followed exactly. Because of substrate neutrality, we can simulate simplified versions of the algorithmic process of evolution in the computer using binary strings of 1s and 0s as the information carriers. These procedures are known as *genetic algorithms* or *evolutionary computation* in computer science.

Cumulative selection will work on almost anything that can yield similar, but non-identical, copies of itself through some replication process. It depends on the presence of a medium that stores information and can be passed on to the next generation; this medium is DNA or RNA (as in certain types of viruses) in terrestrial life forms. Whenever selection acts, it can be thought of as selecting

DNA or RNA copies with particular phenotypic effects over others with different effects. During cumulative *natural selection,* each cycle begins with the results of selection in the previous generation due to the heritability of genetic information.

Most genetic mutations are thought to be either neutral in their phenotypic effects or deleterious, in which case they are removed by *negative selection.* Because of the nonadaptive nature of many mutations, there has also been strong selection for proofreading and error correction mechanisms that reduce the rate at which mutations occur. Rarely, advantageous germline mutations may confer a survival and reproductive advantage on individuals in the next generation who will then, on average, pass on more copies of their genetic material because they will tend to have a larger number of offspring (*positive selection*).

Over many generations, the accumulation of small changes can result in the evolution of DNA (or RNA) sequences with new associated phenotypic effects. This process also gives rise to the evolution of entirely new biological functions. Yet, such advantageous changes are believed to be less frequent than neutral or nearly neutral sequence substitutions at the level of protein evolution. The *neutralist hypothesis* states that therefore most genetic change is not subject to selection, and mutation-driven molecular changes far exceed selection-driven phenotypic changes.

11.1.3.2 Causes of genetic variation

The analysis of evolution at the molecular level must consider the processes which alter DNA sequences. Such alterations are brought about by *mutations* which are errors in DNA replication or DNA repair. These changes provide the genetic variation upon which natural selection can act.

When considered purely as a chemical reaction, complementary base-pairing is not very accurate. It has been estimated that, without the aid of any enzyme, a copied polynucleotide would probably have point mutations at 5-10 positions out of every hundred. Such an error rate of up to 10% would very quickly corrupt a nucleotide sequence. The polymerase enzymes that carry out replication can improve the error rate in two ways—highly accurate nucleotide selection during the polymerization reaction and proofreading carried out by their exonuclease activity.

The genetic material of RNA viruses, including influenza, polio, and HIV viruses, consists of RNA not DNA. These viruses copy themselves by various enzymatic procedures, but they all use a copying polymerase that lacks the more advanced error-correction apparatus. The action of this enzyme is a huge improvement on an enzyme-free copying procedure, but it still has an estimated copying error rate ranging from 1 in 1000 to 1 in 100,000 (10^{-3} to 10^{-5}). A conserved set of proofreading and repair enzymes, absent from RNA viruses, is present in bacteria and all other cellular life, including humans. These new functions, resulting in dramatically reduced copying error rates, made possible the evolution of larger genomes because they could now be copied without incurring a lethal accumulation of mistakes.

Bacteria seem to make copying mistakes at a rate roughly 10,000-1,000,000 times lower than RNA viruses. The bacterium *Escherichia coli* is able to synthesize DNA at an error rate of only 1 in 10^7 nucleotide additions. The overall error rate for replication of the entire *E. coli* genome is only 1 in 10^{10} to 1 in 10^{11} (10^{-10} to 10^{-11}) due to the bacterial mismatch repair system that corrects the errors that the replication enzymes make. The implication is that there will be, on average, only one uncorrected replication error for every 1000 times the *E. coli* genome is copied. In humans, the estimated mutation rate per nucleotide position per generation is 3×10^{-8}, and the total mutation rate per generation is ~200. Various estimates of harmful mutations per generation fall between 2-20. These estimates, approximate as they are, suggest that the error rate is relatively constant between bacteria and humans.

The error-reducing machinery was an extremely valuable evolutionary innovation, but once acquired, it appears to have been either prohibitively difficult or nonadvantageous to optimize it further. In fact, the cellular machinery seems to have evolved to maintain a finely tuned balance between minimizing errors and permitting a certain degree of random mutational change on which natural selection can act. In sexually reproducing organisms, DNA recombination mechanisms for the active generation of genetic variability in gametes exist alongside the error correction machinery. The process of sex cell maturation (meiosis) not only shuffles the maternal and paternal chromosomes by independent assortment, but also promotes the swapping of large pieces of homologous chromosomal stretches.

Mutations do not occur randomly throughout genomes. Some regions are more prone to mutation and are called *hotspots*. In higher eukaryotes, examples of hotspots are short tandem repeats (for nucleotide insertions and deletions) and the dinucleotide 5'-CpG-3' in which the cytosine is frequently methylated and replicated with error, changing it to 5'-TpG-3'. Hotspots in prokaryotes include the dinucleotide 5'-TpT-3' and short palindromes (sequences that read the same on the coding and the complementary strand).

The results of genotypic changes provide the molecular record of evolution. The more closely related two species are, the more similar are their genome sequences and their gene products, i.e. proteins. In the case of proteins, similarity relationships between sequences closely mirror those of their corresponding DNA sequences, but the three-dimensional structures provide more information. In cases where the sequences have changed beyond recognition the protein structures may retain a strong enough resemblance that a potential evolutionary relationship can still be discerned. By systematic analysis of sequences and structures *evolutionary trees* can be constructed which help to reveal the *phylogenetic relationships* between organisms.

11.1.3.3 Classification of mutations

Mutations can be classified by the length of the DNA sequence affected by the mutational event. *Point mutations* (single-nucleotide changes) represent the minimal alteration in the genetic material, but even such a small change can have profound consequences. Some human genetic diseases, such as sickle cell anemia or cystic fibrosis, arise as a result of point mutations in coding regions that cause single amino-acid substitutions in key proteins. Mutations may also involve any larger DNA unit, including pieces of DNA large enough to be visible under the light microscope. Major *chromosomal rearrangements* thus also constitute a form of mutation, though drastic changes of this kind are normally lethal during embryonic development or lead to decreased (zero) fertility. They do therefore contribute only rarely to evolutionary change. In contrast, *silent mutations* result in nucleotide changes that have no effect on the functioning of the genome. Silent mutations include almost all mutations occurring in extragenic DNA and in the noncoding components of genes.

Mutations may also be classified by the type of change caused by the mutational event into *substitutions* (the replacement of one nucleotide by another), *deletions* (the removal of one or more nucleotides from the DNA sequence), *insertions* (addition of one or more nucleotides to the DNA), *inversions* (the rotation by 180 degrees of a double-stranded DNA segment consisting of two or more base pairs), and *recombination*.

11.1.3.4 Substitutional mutation

Some mutations are spontaneous errors in replication which escape the proofreading function of the DNA polymerases that synthesize new polynucleotide chains at the replication fork. This causes a mismatch mutation at positions where the nucleotide that is inserted into the daughter strand is not complementary (by base-pairing) to the corresponding nucleotide in the template DNA. Other mutations arise because of exposure to excessive UV light, highenergy radiation from sources such as X-rays, cosmic rays or radioisotopes, or because a chemical mutagen has reacted with the parent DNA, causing a structural alteration that affects the base-pairing capability of the affected nucleotide. Mutagens include, for example, aflatoxin found in moldy peanuts and grains, the poison gas nitrogen mustard, and polycyclic aromatic hydrocarbons (PAHs) which arise from incomplete combustion of organic matter such as coal, oil, and tobacco. These chemical interactions usually affect only a single strand of the DNA helix, so only one of the daughter molecules carries the mutation. In the next round of replication, it will be passed on to two of the 'granddaughter' molecules. In single-celled organisms, such as bacteria and yeasts, all nonlethal mutations are inherited by daughter cells and become permanently integrated in the lineage that originates from the mutated cell. In multicellular organisms, only mutations that occur in the germ cells are relevant to genome evolution.

Nucleotide substitutions can be classified into *transitions* and *transversions.* Transitions are substitutions between purines (A and G) or pyrimidines (C and T). Transversions are substitutions between a purine and a pyrimidine, e.g., G to C or G to T. The effects of nucleotide substitutions on the translated protein sequence can also form the basis for classification. A substitution is termed *synonymous* if no amino acid change results (e.g., TCT to TCC, both code for serine). *Nonsynonymous,* or amino-acid altering, mutations can be classified into missense or nonsense mutations. A *missense mutation* changes the affected codon into a codon for a different amino acid (e.g., CTT for leucine to CCT for proline); a *nonsense mutation* changes a codon into one of the termination codons (e.g., AAG for lysine to the stop codon TAG), thereby prematurely terminating translation and resulting in the production of a truncated protein. The degeneracy of the genetic code provides some protection against damaging point mutations since many changes in coding DNA do not lead to a change in the amino acid sequence of the encoded protein. Due to the nature of the genetic code, synonymous substitutions occur mainly at the third codon position; nearly 70% of all possible substitutions at the third position are synonymous. In contrast, all the mutations at the second position, and 96% of all possible nucleotide changes at the first position, are nonsynonymous.

11.1.3.5 Insertion and deletion

Replication errors can also lead to insertion and deletion mutations (collectively termed indels). Most commonly, a small number of nucleotides is inserted into the nascent polynucleotide or some nucleotides on the template are not copied. However, the number of nucleotides involved may range from only a few to contiguous stretches of thousands of nucleotides. Replication slippage or slipped-strand mispairing can occur in DNA regions containing contiguous short repeats because of mispairing of neighbouring repeats, and results in deletion or duplication of a short DNA segment. Two important mechanisms- that can give rise to long indels are unequal crossing-over and DNA transposition.

A *frameshift mutation* is a specific type of indel in a coding region which involves a number of nucleotides that is not a multiple of three, and will result in a shift in the reading frame used for translation of the DNA sequence into a polypeptide. A termination codon may thereby be cancelled out or a new stop codon may come into phase with either event resulting in a protein of abnormal length.

11.2 EVOLUTION OF PROTEIN FAMILIES

11.2.1 Protein Families in Eukaryotic Genomes

With the increasing number of fully sequenced genomes, it has become apparent that *multigene families,* groups of genes of identical or similar sequence, are common features of genomes. These are more commonly described as *protein families* by the bioinformatics community. In some cases, the repetition of identical sequences is correlated with the synthesis of increased quantities of a gene product (*dose repetitions*). For example, each of the eukaryotic, and all but the simplest prokaryotic, genomes studied to date contain multiple copies of ribosomal RNAs (rRNAs), components of the protein-synthesizing molecular complexes called ribosomes. The human genome contains approximately 2000 genes for the 5S rRNA (named after its sedimentation coefficient of 5S) in a single cluster on chromosome 1, and roughly 280 copies of a repeat unit made up of 28S, 5.8S and 18S rRNA genes grouped into five clusters on chromosomes 13, 14, 15, 21 and 22.

It is thought that amplification of rRNA genes evolved because of the heavy demand for rRNA synthesis during cell division when thousands of new ribosomes need to be assembled. These rRNA genes are examples of protein families in which all members have identical or nearly identical sequences. Another example of dosage repetition is the amplification of an esterase gene in a Californian *Culex* mosquito resulting in insecticide resistance.

Other protein families, more commonly in higher eukaryotes, contain individual members that are sufficiently different in sequence for their gene products to have distinct properties. Members of some protein families are clustered (*tandemly repeated*), for example the globin genes, but genes belonging to other families are dispersed throughout the genome. Even when dispersed, genes of these families retain sequence similarities that indicate a common evolutionary origin. When sequence comparisons are carried out, it is in certain cases possible to see relationships not only within a single family but also between different families. For example, all of the genes in the α- and β-globin families have discernible sequence similarity and are believed to have evolved from a single ancestral globin gene.

The term *superfamily* was coined by Margaret Dayhoff (1978) in order to distinguish between closely related and distantly related proteins. When duplicate genes diverge strongly from each other in either sequence and/or functional properties, it may not be appropriate to classify them as a single protein family. According to the Dayhoff classification, proteins that exhibit at least 50% amino acid sequence similarity (based on physico-chemical amino acid features) are considered members of a family, while related proteins showing less similarity are classed as belonging to a superfamily and may have quite diverse functions. More recent classification schemes tend to define the cut -off at 35% amino acid identity (rather than using a similarity measure) at the amino acid sequence level for protein families. Following either, definition, the α globins and the β globins are classified as two separate families, and together with the myoglobins form the globin superfamily. Other examples of superfamilies in higher eukaryotes include the collagens, the immunoglobulins and the serine proteases.

One of the most striking examples of the evolution of gene superfamilies by duplication is provided by the homeotic *(Hox)* genes. These genes all share common sequence elements of 180 nucleotides in length, called the homeobox. *Hox* genes playa crucial role in the development of higher organisms as they specify the body plan. Mutations in these genes can transform a body segment into another segment; for example, in *Antennapedia* mutants of *Drosophila* the antennae on the fly's head are transformed into a pair of second legs.

More than 350 homeobox-related genes from animals, plants, and fungi have been sequenced, and their evolution has been studied in great detail as it gives very important clues for understanding the evolution of body plans. *Drosophila* has a single cluster of homeotic genes which consists of eight genes each containing the homeobox element. These eight genes, and other homeobox-containing genes in the *Drosophila* genome, are believed to have arisen by a series of gene duplications that began with an ancestral gene that existed about 1000 million years ago. The modern genes each specify the identity of a different body segment in the fruit fly—a poignant illustration of how gene duplication and diversification can lead to the evolution of complex morphological features in animals.

11.2.2 Gene Duplication

Protein families arise mainly through the mechanism of *gene duplication.* In the 1930s, J.B.S. Haldane and Hermann Muller were the first to recognize the evolutionary significance of this process. They hypothesized that selective constraints will ensure that one of the duplicated DNA segments retains the original, or a very similar, nucleotide sequence, whilst the duplicated (and therefore redundant) DNA segment may acquire divergent mutations as long as these mutations do not adversely affect the organism carrying them. If this is the case, new functional genes and new biological processes may evolve by gene duplication and enable the evolution of more complex organisms from simpler ones. As we have seen in the case of ribosomal RNA, gene duplication is also an important mechanism for generating multiple copies of identical genes to increase the amount of gene product that can be synthesized.

The evolution of new functional genes by duplication is a rare evolutionary event, since the evidence indicates that the great majority of new genes that arise by duplication acquire deleterious mutations. Negative selection reduces the number of deleterious mutations in a population's gene pool. Alternatively, mutations may lead to a version of the copied sequence that is not expressed and does not exert any effect on the organism's phenotype (a silent *pseudogene)*. The most common inactivating mutations are thought to be frameshift and nonsense mutations within the gene-coding region.

Partial gene duplication plays a crucial role in increasing the complexity of genes - segments encoding functional and structural protein domains are frequently duplicated. Domain duplication within a gene may lead to an increased number of active-sites or, analogous to whole gene duplication, the acquisition of a new function by mutation of the redundant segment. In the genomes of eukaryotes, internal duplications of gene segments have occurred frequently. Many complex genes might have evolved from small primordial genes through internal duplication and subsequent modification. The shuffling of domains between different gene sequences can further increase genome complexity.

11.2.3 Mechanisms of Gene Duplication

A duplication event may involve

1. A whole genome (polyploidy)
2. An entire chromosome (aneuploidy or polysomy)
3. Part of a chromosome (partial polysomy)
4. A complete gene
5. Part of a gene (partial or internal gene duplication).

Complete and partial polysomy can probably be discounted as a major cause for gene number expansion during evolution, based on the effects of chromosome duplications in modern organisms. Duplication of individual human chromosomes is either lethal or results in genetic diseases such as Down syndrome, and similar effects have been observed experimentally in *Drosophila*. Whole genome duplication, caused by the formation of diploid rather than haploid gametes during meiosis, provides the most rapid means for increasing gene number. Subsequent fusion of two diploid gametes results in a polyploid genotype, which is not uncommon in plants and may lead to speciation. A genome duplication event has also been identified in the evolutionary history of the yeast *Saccharomyces cerevisiae* and has been dated to have taken place approximately 100 million years ago. Genome duplication provides the potential for the evolution of new genes because the extra genes can undergo mutational changes without harming the viability of the organism.

Complete or partial gene (domain) *duplication* can occur by several mechanisms. *Unequal crossing-over* during meiosis is a reciprocal *recombination* event that creates a sequence duplication in one chromosome and a corresponding deletion in the other. It is initiated by similar nucleotide sequences that are not at identical locations in a pair of homologous chromosomes and cause the chromosomes to align to each other out of register. *Unequal sister chromatid exchange* occurs by the same mechanism, but involves a pair of chromatids from the same chromosome. *Transposition* is defined as the movement of genetic material *(transposons)* from one chromosomal location to another. Transposons carry one or more genes that encode other functions in addition to those related to transposition.

There are two types of transposition events. In conservative transposition the DNA element itself moves from one site to another, whilst in replicative transposition, the element is copied and one copy remains at its original site while the other inserts at a new site. Replicative transposition therefore involves an increase in the copy number of the transposon. Some transposable elements transpose via a RNA intermediate. All such elements contain a reverse transcriptase gene whose product is responsible for the relatively error-prone reverse transcription of the DNA element.

Transposable elements can promote gene duplications, and genomic rearrangements including inversions, translocations, duplications and large deletions and insertions. For example, a duplication of the entire growth hormone gene early in human evolution might have occurred via *Alu-Alu* recombination *(Alu* repeats represent one of the most abundant families of retrosequences in the human genome).

11.2.4 Concept of Homology

Evolutionary relationships between members of protein families can be classified according to the concept of *homology*. Sequences or structures are homologous if they are related by evolutionary divergence from a common ancestor. From this follows that homology cannot be directly observed, but must be inferred from calculated levels of sequence or structural similarity. Reliable threshold values of similarity are dependent on the mathematical methods used for analysis (this will be discussed in detail in later chapters). In contrast, analogy refers to the acquisition of a shared feature (protein fold or function) by convergent evolution from unrelated ancestors.

Among homologous sequences or structures, it is possible to distinguish between those that have resulted from gene duplication events within a species genome and perform different but related functions within the same organism (*paralogs*) and those that perform the same, or a highly similar, function in different species (*orthologs*). Horizontal gene transfer is defined as the transfer of genetic material from one genome to another, specifically between two species. Sequences acquired via horizontal gene transfer are called *xenologs*.

A high degree of gene conservation across organisms, correlated with conservation of function, has been well established by whole genome sequencing. For example, comparison between two complete eukaryotic genomes, the budding yeast *Saccharomyces cerevisiae* and the nematode worm *Caenorhabditis elegans*, suggested *orthologous relationships* between a substantial number of genes in the two organisms. About 27% of the yeast genome (~5700 genes) encode proteins with significant similarity to ~12% of nematode genes (~18,000). Furthermore, the same set of yeast genes also has putative orthologs in the *Drosophila* genome. Most of these shared proteins perform conserved roles in core biological processes common to all eukaryotes, such as metabolism, gene transcription, and DNA replication.

11.2.5 Modularity of Proteins

Domain shuffling is a term applied to both domain duplication within the same gene sequence and domain insertion. *Domain insertion* refers to the exchange of structural or functional domains between coding regions of genes or insertion of domains from one region into another. Domain insertion results in the generation of *mosaic* (or *chimeric*) *proteins.* Domain insertion may be preceded by domain duplication events.

One of the first mosaic proteins to be studied in detail was *tissue plasminogen activator* (*TPA*). TPA converts plasminogen into its active form, plasmin, which dissolves fibrin in blood clots. The TPA gene contains four well-characterized DNA segments from other genes: a fibronectin finger domain from fibronectin or a related gene; a growth-factor module that is homologous to that present in epidermal growth factor and blood clotting enzymes; a carboxy-terminal region homologous to the proteinase modules of trypsin-like serine proteinases; and two segments similar to the kringle domain of plasminogen. A special feature of the organization of the TPA gene is that the junctions between acquired domains coincide precisely with exon-intron borders. The evolution of this composite gene thus appears to be a likely result of exon shuffling. The domains present in TPA can also be identified in various arrangements in other proteins.

The combination of multiple identical domains, or of different domains, to form mosaics gives rise to enormous functional and structural complexity in proteins. Because domains in mosaics perform context-dependent roles and related modules in different proteins do not always perform exactly the same function, the domain composition of a protein may tell us only a limited amount about the integrated function of the whole molecule.

11.3 EVOLUTION TAKES PLACE AT ALL LEVELS OF BIOLOGICAL ORGANIZATION

Contemporary evolutionary thought is complementary to the study of gene and protein evolution through bioinformatics. This intellectually engaging and lively area of research continues to pose fundamental challenges to our understanding of the mechanisms by which natural selection exerts its effects on genomes. Whilst all evolutionary change can be represented as a change in the frequencies of particular DNA segments within genomes, an ongoing debate is concerned with the question of *why* these frequencies change. The received view conceives of natural selection as the result of competition between individual organisms in a population, and by default, their individual copies of the species genome.

Gene selectionism fakes the view that only relatively short segments of DNA, able to confer certain phenotypic effects on the organism, satisfy a crucial requirement for units of selection. According to this view, the natural selection of individual phenotypes cannot by itself produce cumulative change, because phenotypes are extremely temporary manifestations, and genomes associated with particular phenotypes are broken up when gametes are formed, In contrast, genes in germline cell lineages are passed on intact through many generations of organisms and are thus exposed to cumulative selection.

This argument can even be taken a step further. Extensive sequence comparisons, made possible by the large-scale sequencing of prokaryotic, archeal and eukaryotic genomes, have consistently revealed the modular composition of genes and proteins, and have enabled detailed evolutionary studies of their constituent smaller units, protein *domains*. Domains are discrete molecular entities that form the stable components of larger genes and proteins and are passed on intact between generations. Thus, DNA segments encoding protein domains can arguably be seen as stable units of selection.

One preliminary conclusion that can be drawn from this debate is that a range of mechanisms is likely to be operative in genome evolution. Bioinformatics can analyze the resulting patterns of evolutionary change present in biological sequences and three-dimensional protein structures. In elucidating the complex underlying processes, other fields of biology, notably molecular biology and evolutionary biology, are bioinformatics' indispensable partners.